OUTLINES

of

PSYCHOLOGY

Wilhelm Max Wundt

Translated by Charles Hubbard Judd

Printed in Scotts Valley, CA - USA.

Wundt, Wilhelm Max.

Outlines of Psychology / **Wilhelm Max Wundt** – 1st ed.

1. Psychology

Table of Contents

INTRODUCTION

CHAPTER 1. PSYCHICAL ELEMENTS

CHAPTER 2. PSYCHICAL COMPOUNDS

CHAPTER 3. INTERCONNECTIONS OF PSYCHICAL COMPOUNDS

CHAPTER 4. PSYCHICAL DEVELOPMENTS

CHAPTER 5. PSYCHICAL CAUSALITY AND ITS LAWS

INTRODUCTION

1. PROBLEM OF PSYCHOLOGY

1. Two definitions of psychology have been the most prominent in the history of this science. According to one, psychology is the "science of mind", psychical processes being regarded as phenomena from, which it is possible to infer the nature of an underlying metaphysical mind-substance. According to the other, psychology is the "science of inner experience"; psychical processes are here looked upon as belonging to a specific form of experience, which is readily distinguished by the fact that its contents are known through "introspection", or through the "inner sense" as it is called if one uses the phrase which has been employed to distinguish introspection from sense-perception through the outer senses.

Neither of these definitions, however, is satisfactory to the psychology of today. The first or metaphysical definition belongs to a period of development that lasted longer in this science than in others. But is here, too, forever left behind, since psychology has developed into an empirical discipline, operating with methods of its own; and since the "mental sciences" have gained recognition as a great department of scientific investigation, distinct from the sphere the natural sciences, and requiring as a general groundwork an independent psychology, free from all metaphysical theories.

The second or empirical definition, which sees in psychology a "science of inner experience", is inadequate because it may give rise to the misunderstanding that psychology has to do with objects totally different from the objects of so called "outer experience". It is, indeed, true that there are certain contents of experience which belong in the sphere of psychological investigation, and are not to be found among the objects and processes studied by natural science; such are our feelings, emotions, and decisions. On the other hand, there is not a single natural phenomenon that may not, from a different point of view, become an object of psychology. A stone, a plant, a tone, a ray of light, are, when treated as natural phenomena, objects of mineralogy, botany, physics, etc. In so far, however, as they are at the same time *ideas,* they are objects of psychology, for psychology seeks to account for the genesis of these ideas, and for their relations, both to other ideas and to those psychical processes, such as feelings, volitions, etc., which are not referred to external objects. There is then, no such thing as an "inner sense" which can be regarded as an organ of introspection, and as distinct from the outer senses, or organs of objective perception. The ideas of which psychology seeks to investigate the attributes, are identical with those upon which natural science is based; while the subjective activities of feeling, emotion, and volition, which are neglected in natural science, are not known through special organs but are directly and inseparably connected with the ides referred to external objects.

2. It follows, then, that the expressions outer and inner experience do not indicate different objects, but *different points of view* from which we take up the consideration and scientific treatment of a unitary experience. We are naturally led to these points of view, because every concrete experience immediately divides into *two factors:* into a *content* presented to us, and our *apprehension* of this *content.* We call the first of these factors *objects of experience,* the second, *experiencing subject.* This division indicates two directions for the treatment of experience. One is that of the natural sciences, which concern themselves with the *objects* of experience, thought of as independent of the subject. The other is that of *psychology,* which investigates the whole content of experience in its relations to the subject and also in regard to the attributes which this content derives directly from the subject. The point of view of natural science may, accordingly, be designated as that of mediate experience, since it is possible only after abstracting from the subjective factor present in all actual experience; the point of view of psychology, on the other hand, may be designated as that of *immediate experience,* since it purposely does away with this abstraction and all its consequences.

3. The assignment of this problem to psychology, making it a general, empirical science coordinate with the natural sciences, and supplementary to them, is justified by the method of all the *mental sciences,* for which psychology furnishes the basis. All of these sciences, philology, history and political and social science, have as their subject-matter, immediate experience as determined by the interaction of objects with knowing and acting subjects. None of the mental sciences employs the abstractions and hypothetical supplementary concepts of natural science; quite otherwise, they all accept ideas and the accompanying subjective activities as immediate reality. The effort is then made to explain the single components of this reality through their mutual interconnections. This method of psychological interpretation employed in *each of the special mental sciences,* must also be the mode of procedure in psychology itself, being the method required by the subject-matter of psychology, the immediate reality of experience.

Since natural science investigates the content of experience after abstracting from the experiencing subject, its problem is usually stated as that of acquiring "knowledge of the outer world". By the expression outer world is meant the sum total of all the objects presented in experience. The problem of psychology has sometimes been correspondingly defined as "self knowledge of the subject". This definition is, however, inadequate, because the interaction of the subject with the outer world and with other similar subjects is just as much a part of the problem of psychology as are the attributes of the single subject. Furthermore, the expression can easily be interpreted to mean that the outer world and the subject are separate components of experience, or, at least, components which can be distinguished as independent contents of experience, whereas, in truth, outer experience is always connected with the apprehending and knowing functions of the subject, and inner experience always contains ideas from the outer world as indispensable components. This interconnection is the necessary result of the fact that in reality experience is not a mere juxtaposition of different elements, but a single organized whole which requires in each of its components the subject which apprehends the content, and the objects which are presented as content. For this reason natural science can not abstract from the knowing subject entirely, but only from those attributes of the subject which either disappear entirely when we remove the subject in thought, as, for example, the feelings, or from those attributes which must be regarded on the ground of physical researches as belonging to the subject, as, for example, the qualities of sensations. Psychology, on the contrary, has as its subject of treatment the *total*, content of, experience in its immediate character.

The only ground, then, for the division between natural, science on the one hand, and psychology and the mental sciences on the other, is to be found in the fact that all in the fact that all experience contains as its factors a content objectively presented, and an experiencing subject. Still, it is by no means necessary that *logical* definitions of these two factors should precede the separation of the sciences from one another, for it is obvious that such definitions are possible only after they have a basis in the investigations of natural science and of psychology. All that it is necessary to presuppose from the first is the consciousness which accompanies all experience, that in this experience objects are being presented to a subject. There can be no assumption knowledge of the conditions upon which the distinction is based, or of the definite characteristics by which one factor is to be distinguished from the other. Even the use of the terms object and subject in this connection must be regarded as the application to the first stage of experience, of distinctions which are reached only through developed logical reflection.

The forms of interpretation in natural science and psychology are supplementary, not only in the sense that the first considers objects after abstracting, as far as possible, from the subject, while the second has to do with the part which the subject plays in the rise of experience; but they are also supplementary in the sense that each takes a different point of view in considering any single content of experience. Natural science seeks to discover the nature of objects without reference to the subject. The knowledge that it produces is therefore *mediate* or *conceptual*. In place of the immediate objects of experience, it sets concepts gained from these objects by abstracting from the subjective components of our ideas. This abstraction makes it necessary continually to supplement reality with hypothetical elements. Scientific analysis shows that many components of experience ~ as, for example, sensations ~ are subjective effects of objective processes. These objective processes in their objective character, independent of the subject, can therefore never be a part of experience. Science makes up for this lack of direct contact with the objective processes, by forming supplementary hypothetical concepts of the objective properties of matter. Psychology, on the other hand, investigates the contents of experience in their complete and actual form, both the ideas that are referred to objects, and also the subjective processes which cluster about these ideas. The knowledge thus gained in psychology is, therefore, *immediate* and *perceptual*, ~ perceptual in the broad sense of the term in which, not only sense-perceptions, but all *concrete reality* is distinguished from all that is abstract and conceptual in thought. Psychology can exhibit the interconnection of the contents of experience, as these interconnections are actually presented to the subject, only by avoiding entirely the abstractions and supplementary concepts of natural science. Thus, while natural science and psychology are both empirical sciences in the sense that they aim to explain the contents of experience, though from different points of view, it is obvious that, in consequence of the special character of its problem, psychology must be recognized as the *more strictly empirical.*

2. GENERAL THEORIES OF PSYCHOLOGY

1. The view that psychology is an empirical science which deals, not with a limited group of specific contents of experience, but with the immediate contents of all experience, is of recent origin. It encounters even in the science of today hostile views, which are to be looked upon, in general, as the survivals of earlier stages of development, and which are in turn arrayed against one another according to their attitudes on the question of the relations of psychology to philosophy and to the other sciences. On the basis of the two definitions mentioned above (sec.1, 1) as being the most widely accepted, two chief forms of psychology may be distinguished: *metaphysical psychology* and *empirical psychology.* Each is further divided into a number of special tendencies.

Metaphysical psychology generally values very little the empirical analysis and causal interpretation of psychical processes. Regarding psychology as a part of philosophical metaphysics, the chief effort of such psychology is directed toward the discovery of a definition of the "nature of mind" which shall be in accord with the metaphysical system to which the particular form of psychology belongs. After a metaphysical concept of mind has thus been established, the attempt is made to deduce from it the actual content of psychical experience. The characteristic which distinguishes metaphysical psychology from empirical psychology, then, is its attempt to deduce psychical processes, not from other psychical processes, but from some substratum entirely unlike these processes themselves: either from the manifestations of a special mind-substance, or from the attributes and processes of matter. At this point metaphysical psychology branches off in *two* directions. *Spiritualistic psychology* considers psychical the manifestations of a *specific* mind-substance, which is regarded either as essentially different form matter (dualism), or as related in nature to matter (*monism* or *monadalogy*). The fundamental metaphysical doctrine of spiritualistic psychology is the assumption of the *supersensible* nature of mind, and in connection with this, the assumption of its immortality. Sometimes the further notion of preexistence is also added. *Materialistic psychology,* on the other hand refers psychical processes to the *same* material substratum as that which natural science employs for the hypothetical explanation of natural phenomena. According to this view, psychical processes, like physical vital processes, are connected with certain organizations of material particles which are formed during the life of the individual and broken up at the end of that life. The metaphysical character of this form of psychology is determined by its denial that the mind is supersensible in its nature as is asserted by spiritualistic psychology. Both theories have this in common, that they seek not to interpret psychical experience from experience itself, but to derive it from presuppositions about hypothetical processes in a metaphysical substratum.

2. From the strife that followed these attempts at metaphysical explanation, *empirical psychology* arose. Wherever empirical psychology is consistently carried out, it strives either to arrange psychical processes under general concepts derived directly from the interconnection of these processes themselves, or it begins with certain, as a rule simpler processes, and then explains the more complicated as the result of the interaction of those with which it started. There may be various fundamental principles for such an empirical interpretation, and thus it becomes possible to distinguish several varieties of empirical psychology. In general, these may be classified according to *two* principles of division. The first has reference to the relation of inner and outer experience and to the attitude which the two empirical sciences, natural science and psychology, take toward each other. The second had reference to the facts or concepts derived from these facts, which are used for the interpretation of psychical processes. Every system of empirical psychology takes its place under both of these principles of classification.

3. On the *general question as to the nature of psychical experience* the two views already mentioned. (sec. 1) on account of their decisive significance in determining the problem of psychology: *psychology of the inner sense, and psychology as the science of immediate experience.* The first treats psychical processes as contents of a sphere of experience coordinate with the sphere of experiences which, derived through the outer senses, is assigned as the province of the natural sciences, but though coordinate is totally different from it. The second recognizes no real difference between inner and outer experience, but finds the distinction only in the different *points of view* from which unitary experience is considered in the two cases.

The first of these two varieties of empirical psychology is the older. It arose primarily through the effort to establish the independence of psychical observation in opposition to the encroachments of natural philosophy. In thus coordinating natural science and psychology, it sees the justification for the equal recognition of both spheres of science in the fact that they have entirely different objects and modes of perceiving these objects. This view has influenced empirical psychology in two ways. First, it favored the opinion that psychology should employ empirical methods, but that these methods, like psychological experience, should be fundamentally different from those of natural science. Secondly, it gave rise to the necessity of showing some connection or other between these two kinds of experience, which were supposed to be different. In regard to the first demand, it was chiefly the psychology of the inner sense that developed the method of *pure introspection* (sec. 3, 2). In attempting to solve the second problem, this psychology was necessarily driven back to a metaphysical basis, because of its assumption of a difference between the physical and the psychical contents of experience. For, from the very nature of the case, it is impossible, to account for the relations of inner to outer experience, or the so-called "interaction between body and mind", from the position here taken, except through metaphysical presuppositions. These presuppositions must then, in turn, affect the psychological investigation itself in such a way as to result in the importation of metaphysical hypotheses into it.

4. Essentially distinct from the psychology of the inner sense is the form of psychology which defines itself as "the science of immediate experience". Regarding, as it does, outer and inner experience, not as different parts of experience, but as different ways of looking at one and the same experience, this form of psychology can not admit any fundamental difference between the methods of psychology and those of natural science. It has, therefore, sought above all to cultivate *experimental* methods which shall lead to just such an exact analysis of psychical processes as that which the explanatory natural sciences undertake in the case of natural phenomena, the only differences being those which arise from the diverse points of view. It holds, also, that the special mental sciences which have to do with concrete mental processes and creations, stand on the same basis of a scientific consideration of the immediate contents of experience and of their relations to acting subjects. It follows, then, that psychological analysis of the most general mental products, such as language, mythological ideas, and laws of custom, is to be regarded as an aid to the understanding of all the more complicated psychical processes. In its methods, accordingly, this form of psychology stands in close relation to other sciences: as *experimental* psychology, to the natural sciences; as *social psychology*, to the special mental sciences.

Finally, from this point of view, the question of the relation between psychical and physical objects disappears entirely. They are not different objects at all, but one and the same content of experience, looked at in one case - that of the natural sciences - after abstracting from the subject, in the other - that of psychology - in their immediate character and complete relation to the subject. All metaphysical hypotheses as to the relation of psychical and physical objects are, when viewed from this position, attempts to solve a problem which never would have existed if the case had been correctly stated. Though psychology must then dispense with metaphysical supplementary hypotheses in regard to the interconnection of psychical processes, because these processes are the immediate contents of experience, still another method of procedure, however, is open since inner and outer experience are supplementary points of view. Wherever breaks appear in the interconnection of psychical processes, it is allowable to carry on the investigation according to the physical methods of considering these same processes, in order to discover whether the absent link can be thus supplied. The same holds for the reverse method of filling up the breaks in the continuity of our physiological knowledge, by means of elements derived from psychological investigation. Only on the basis of such a view, which sets the two forms of knowledge in their true relation, is it, possible for psychology to become in the fullest sense an empirical science. Only in this way, too, can physiology become the true supplementary science of psychology, and psychology, on the other hand, the auxiliary of physiology.

5. Under the *second* principle of classification mentioned above (2), that is, according to the *facts or concepts with which the investigation of psychical processes starts*, there are *two* varieties of empirical psychology to be distinguished. They are, at the same time,

successive stages in the development of psychological interpretation. The first corresponds to a *descriptive*, the second to an *explanatory* stage. The attempt to present a discriminating description of the different psychical processes, gave rise to the need of an appropriate *classification*. Class concepts were formed, under which the various processes were grouped; and the attempt was made to satisfy the need of an interpretation in each particular case, by subsuming the components of a given compound process under their proper class concepts. Such concepts are, for example, sensation, knowledge, attention, memory, imagination, understanding, and will. They correspond to the general concepts of physics which are derived from the immediate perception of natural phenomena, such as weight, heat, sound, and light. Like those concepts of physics, these derived psychical concepts may serve for a first grouping of the facts, but they contribute nothing whatever to the explanation of these facts. Empirical psychology has, however, often been guilty of confounding this description with explanation. Thus, the *faculty-psychology* considered these class-concepts as psychical forces or faculties, and referred psychical processes to their alternating or united activity.

6. Opposed to this method of treatment found in descriptive faculty-psychology, is that of *explanatory* psychology. When consistently empirical, the latter must base its interpretations on certain facts which themselves belong to psychical experience. These facts may, however, be taken from different spheres of psychical activity, and so it comes that explanatory treatment may be further divided into *two* varieties which correspond respectively to the two factors, objects and subject, which go to make up immediate experience. When the chief emphasis is laid on the *objects* of immediate experience, *intellectualistic psychology*, this type of psychology attempts to derive all psychical processes, especially the subjective feelings, impulses, and volitions, from *ideas*, or *intellectual processes* as they may be called on account of their importance for knowledge of the objective world. If, on the contrary, the chief emphasis is laid on the way in which immediate experience arises in the subject, a variety of explanatory psychology results which attributes to those subjective activities referred to external objects, a position as independent as that assigned to ideas. This variety has been called *voluntaristic psychology*, because of the importance that must be conceded to volitional processes in comparison with other subjective processes.

Of the two varieties of psychology which result from the general attitudes on the question of the nature of inner experience (3), psychology of the inner sense commonly tends towards intellectualism. This is due to the fact that, when the inner sense is coordinated with the outer senses, the contents of psychical experience which first attract consideration are those which are presented as objects to this inner sense in a manner analogous to the presentation of natural objects to the outer senses. It is assumed that the character of objects can be attributed to *ideas* alone of all the contents of psychical experience, because they are regarded as *images* of the external objects presented to the outer senses. Ideas are, accordingly, looked upon as the only real objects of the inner sense while all processes not referred to external objects, as, for example, the feelings, are interpreted as obscure ideas, or as ideas related to one's own body, or, finally, as effects arising from combinations of ideas.

11

The psychology of immediate experience (4), on the other hand, tends toward voluntarism. It is obvious that here, where the chief problem of psychology is held to be the investigation of the subjective rise of all experience, special attention will be devoted to those factors from which natural science abstracts.

7. *Intellectualistic* psychology has in the course of its development separated into *two* trends. In one, the *logical* processes of judgment and reasoning are regarded as the typical forms of all psychoses; in the other, certain combinations of successive memory-images distinguished by their frequency, the so-called *associations of ideas*, are accepted as such. The *logical theory* is most clearly related to the popular method of psychological interpretation and is, therefore, the older. It still finds some acceptance, however, even in modern times. The *association-theory* arose from the philosophical empiricism of the last century. The two theories stand to a certain extent, in antithesis, since the first attempts to reduce the totality of psychical processes to higher, while the latter seeks to reduce it to the lower and, as it is assumed, simpler forms of intellectual activity. Both are one-sided, and not only fail to explain affective processes and volitional processes on the basis of the assumption with which they start, but are not able to give a complete interpretation even of the intellectual processes.

8. The union of psychology of the inner sense with the intellectualistic view has led to a peculiar assumption that has been in many cases fatal to psychological theory. We may define this assumption briefly as the *erroneous attribution of the nature of things to ideas, to ideas.* Not only was an analogy assumed between the objects of so-called inner sense and those of the outer senses, but former were regarded as the images of the latter; it came that the attributes which natural science ascribes to external objects, were also transferred to the immediate objects of the "inner sense", the ideas. The assumption was made that ideas are themselves things, just as the external objects to which we refer them; that they disappear from consciousness and come back into it; that they may, indeed, be more or less intensely and clearly perceived, according as the inner sense is stimulated through the outer senses or not, and according to the degree of attention concentrated upon them, but that on the they remain unchanged in qualitative character.

9. In all these respects *voluntaristic psychology* is opposed to intellectualism. While the latter assumes an inner sense and specific objects of inner experience, volunteerism is closely related to the view that inner experience is identical with *immediate* experience. According to this doctrine, the content psychological experience does not consist of a sum of objects, but of all that which makes up the process of experience in general, that is of all the experiences of the subject in their immediate character, unmodified by abstraction or reflection. It follows of necessity that the contents of psychological experience are here regarded as an *interconnection of processes.*

This concept of *process* excludes the attribution of an objective and more or less permanent character to the contents of psychical experience. Psychical facts are *occurrences*, not objects; they take place, like all occurrences, in time and are never the same at a given point in time as they were during the preceding moment. In this sense *volitions* are *typical* for all psychical processes. Voluntaristic psychology does not by any means assert that volition is the only real form of psychosis, but merely that, with its closely related feelings and emotions, it is just as essential a component of psychological experience as sensations and ideas. It holds, further, that all other psychical processes are to be thought of after the analogy of volitions, they too being a series of continuous changes in time, not a sum of permanent objects, as intellectualism generally assumes in consequence of its erroneous attribution to ideas of those properties which we attribute to external objects. The recognition of the *immediate* reality of psychological experience excludes the possibility of the attempt to derive the particular components of psychical phenomena from any others specifically different. The analogous attempts of metaphysical psychology to reduce all psychological experience to the heterogeneous, imaginary processes of a hypothetical substratum are, for the same reason, inconsistent with the real problem of psychology. While it concerns itself, however, with immediate experience, psychology assumes from the first that all psychical contents contain objective as well a subjective factors. These are to be distinguished only through deliberate abstraction, and can never appear as really separate processes. In fact, immediate experience shows that there are no ideas which do not arouse in us feelings and impulses of different intensities, and, on the other hand, that a feeling or volition is impossible which does not refer to some ideated object.

10. The governing principles of the psychological position maintained in the following chapters may be summed up in *three* general statements.

1) Inner, or psychological experience is not a special sphere of experience apart from others, but is *immediate experience* in its totality.

2) This immediate experience is not made up of unchanging contents but of an *interconnection of processes*, not of objects, but of *occurrences*, of *universal human experiences* and their relations in accordance with certain laws.

3) Each of these processes contains an *objective content* and a *subjective process*, thus including the general conditions both of all knowledge and of all practical human activity.

Corresponding to these three general principles, we have a *threefold relation of psychology* to the other sciences.

1) As the science of immediate experience, it is *supplementary* to the *natural sciences*, which, in consequence of their abstraction from the subject, have to do only with the objective, *mediate* contents of experience. Any particular fact can, strictly speaking, be understood in its full significance only after it has been subjected to the analyses of both natural science and psychology. In this sense, then, physics and physiology are auxiliary to psychology, and the latter is, in turn, supplementary to the natural sciences.

2) As the science of the universal forms of immediate human experience and their combination in accordance with certain laws, it is the *foundation of the mental sciences*. The subject-matter of these sciences is in all cases of the activities proceeding from immediate human experiences, and their effects. Since psychology has for its problem the investigation of the forms and laws of these activities, it is at once the most, general mental science, and the foundation for all the others, such as philology, history, political economy, jurisprudence, etc.

3) Since psychology pays equal attention to *both* the subjective and objective conditions which underlie not only theoretical knowledge, but practical activity as well, and since it seeks to determine their interrelation, it is the empirical discipline whose results are most immediately useful in the invention of the general problems of the *theory of knowledge*, and *ethics*, the two foundations of *philosophy*. Thus, psychology is, in relation to the natural sciences, the *supplementary*, in relation to the mental sciences the *fundamental*, and in relation to philosophy it is the *propaedeutic empirical science.*

10a. The view that it is not a difference in the objects of experience, but in the way of treating experience, that distinguishes psychology from natural science has come to be recognized more and more in modern psychology. Still a clear comprehension of the essential character of this position in regard to the scientific problems of psychology, is prevented by the persistence of older tendencies derived from metaphysics and natural philosophy. Instead of starting from the fact that the natural sciences are possible only after abstracting from the subjective factors of experience, the more general problem of treating the contents of *all* experience in the most general way, is sometimes assigned to natural science. In such a case psychology is, of course, no longer coordinate with the natural sciences, but subordinate to them. Its problem is no longer to remove the abstraction employed by the natural sciences, and in this way to gain with them a complete view of experience, but it has to use the concept "subject" furnished by the natural sciences, and to give an account of the influence of this subject on the contents of experience. Instead of recognizing that an adequate definition of "subject" is possible only as a result of psychological investigations (sec. 1, 3a), a finished concept formed exclusively by the natural sciences is here foisted upon psychology. Now. for the natural sciences the subject identical with the body. Psychology is accordingly defined as the science which has to determine the dependence of immediate experience on the body. This position, which may be designated "psycho-physical materialism", is epistemologically untenable and psychologically unproductive. Natural science, which purposely abstracts from the subjective component of all experience, is at least in a position to give a final definition of the subject. A psychology that starts with such a purely physiological definition depends, therefore, not on experience but, just like the older materialistic psychology, on a metaphysical presupposition. The position is psychologically unproductive because, from the very first, it turns over the causal interpretation of psychical processes to physiology. But physiology has not yet furnished such an interpretation and never will be able to do so, because of the difference between the manner of regarding phenomena in natural science and in psychology. It is obvious, too, that such a form of psychology, which been turned into hypothetical brain-mechanics, con never be of any service as a basis for the mental sciences.

14

The *strictly empirical* trend of psychology, defined in the principles formulated above, is opposed to these attempts to renew metaphysical doctrines. In calling it "voluntaristic", we are not to overlook the fact that, in itself, this psychological voluntarism has absolutely no connection with any metaphysical doctrine of will. Indeed it stands in opposition to Schopenhauer's one-sided metaphysical voluntarism, which derived all from being from a transcendental original will, and to the metaphysical systems of a Spinoza or a Herbart, which arose from intellectualism. In its relation to metaphysics, the characteristic of psychological voluntarism in the sense above defined, is its exclusion of all metaphysics from psychology. In its relations to other forms of psychology, it refuses to accept any of the attempts to reduce volitions to mere ideas, and at the same time emphasizes the *typical* character of volition for all psychological experience. Volitional acts are universally recognized as *occurrences*, made up of a series of continual changes in quality and intensity. They are typical in the sense that this characteristic of being occurrences is held to he true for all the contents of psychical experience.[1]

---

[1] *Classics* Editor's note: In the 3rd English edition of *Outlines* (Judd, Trans, 1907), Wundt included the following short essays about sources for the various forms of psychology reviewed in this Introduction.

In their historical development many of these forms of psychology have grown up together. One may, however, mark off certain general sequences. Thus, metaphysical forms have generally preceded empirical forms; descriptive forms have preceded explanatory; and finally, intellectualism has preceded voluntarism. The oldest work which treated of psychology as an independent science was ARISTOTLE'S work entitled "On the Soul". This work is to be classified as belonging to the dualistic group in its metaphysics, and to the group of faculty-psychologies on the side of its empirical explanations. (The soul was treated as the living principle in the body. There were three fundamental faculties, namely, alimentation, sensation, and thought,) Modern spiritualistic psychology begins with DESCARTES' dualism which recognizes two distinct forms of reality: first, the soul as a thinking and unextended entity, and second, matter as an extended and non-thinking reality. The Cartesian system found the point of contact between these two forms of reality in a particular region of the human brain, namely, the, pineal gland. The founder of modern materialism is THOMAS HOBBES (1588-1679). (The ancient materialistic dualism of DEMOCRATES had not yet differentiated itself from spiritualistic dualism). HOBBES, together with LA METTRIE and HOLBACH developed in the 18th century a mechanical materialism, while DIDEROT and HELVETIUS developed a psycho-physical materialism which has representatives even in present times. Spiritualistic monism first arose in the monadology of LEIBNIZ. In modern times this has been taken up by HERBART and his school, by LOTZE, and others. The establishment of the psychology of the inner sense may be properly attributed to JOHN LOCKE (1632-1704). This form of psychology has been defended in modern times, to some extent by KANT, and with special emphasis by EDUARD BENEKE, (1798-1854), K. FORTLAGE, and others. Modern faculty-psychology arose with the work of CHRISTIAN WOLFF (1679-1754), who distinguished as the chief faculties, knowledge and desire. Since the time of TETENS (1736-1805) three faculties have been more commonly accepted than WOLFF'S two. PLATO named these three, as did also KANT. They are knowledge, feeling and desire. Logical intellectualism is the oldest of the explanatory forms of psychology. This corresponds directly to the popular interpretation of psychical processes. The earlier empiricists, as for example LOCKE, and even BERKELEY (16481753) who in his "Essay towards a New Theory of Vision" anticipates modern experimental psychology, are to be classed as representatives of logical intellectualism. This view is at the present time to be found in the psychological discussions indulged in by physiological writers, when they treat of such topics as sense perception. Among the philosophical representatives of this logical intellectualism in our day, one must mention especially FRANZ BRENTANO and his school. Association psychology is first found in the works of two writers who appear at about the same time, namely, DAVID HARTLEY (1704-1757) and DAVID

HUME (1711-1776). These two writers represent, however, two different tendencies which continue even in present-day psychology. HARTLEY's association psychology refers the association processes to certain physiological conditions, while HUME's regards the association process as a psychological process. The first form allies itself, accordingly, to psycho-physical materialism; this is found in the works of such a modern writer as HERBERT SPENCER. Closely related to HUME's psychological associationism is the psychology of HERBART. HERBART's doctrine of the statics and mechanics of ideas is a purely intellectualistic doctrine. (Feeling and volition are here recognized only as certain phases of ideas). It is in agreement with associationism in its fundamental mechanical view of mental life. This similarity is not to be overlooked merely because Herbart sought through certain hypothetical assumptions to give his psychological discussions an exact mathematical form. There are many anticipations of voluntaristic psychology in the works of psychologists of the "pure introspection" school, and of the association schools. The first thorough-going exposition of this form of psychology was the work of the author of this *Outlines of Psychology* in his psychological treatises. It is to be noted that this psychological voluntarism, as, indeed, one can see from the description which has already been given, is to be clearly distinguished from metaphysical voluntarism as developed by such a writer as SCHOPENHAUER. Metaphysical voluntarism seeks to reduce everything to an original transcendental will, which lies back of the phenomenal world and serves as a substratum for this world. Psychological voluntarism on the other hand, looks upon empirical volitional processes with their constituent feelings, sensations, and ideas, as the types of all conscious processes. For such a voluntarism even volition is a complex phenomenon which owes its typical significance to this very fact that it includes in itself the different kinds of psychical elements.

References. Psychology of the inner sense: LOCKE, An Essay concerning Human Understanding, 1690. EDUARD BENEKE, Psychologische Skizzen, 2 vols., 1825-1827, and Lehrbuch der Psychologie als Naturwissenschaft, 1833, 4th ed. 1877. K. FORTLAGE, System der Psychologie, 2 Vols., 1855. Faculty-psychology: CHRISTIAN WOLFF, Psychologia empirica, 1732, Psychologia rationalis, 1734; and Vernunftige Gedanken von Gott, der Welt, der Seele des Menschen etc., 1719. TETENS, Philosophische Versuche uber die menschliche Natur, 1776-1777. KANT, Anthropologie, 1798 (a practical psychology, well worth reading even at this late date because of its many nice observations).

Association psychology: HARTLEY Observations on Man, his Frame, his Duties, his Expectations, 1749. PRIESTLY, Hartley's Theory of the Human Mind on the Principles of the Association of Ideas, 1775. HUME, Treatise on Human Nature, 1734 -1737; and Enquiry concerning Human Understanding, 1748. JAMES MILL, Analysis of the Phenomena of the Human Mind, 1829, later edited with notes by Alexander Bain, John Stuart Mill and others, 2nd ed. 1878. ALEXANDER BAIN, The Senses and the Intellect, 1855, 4th ed. 1894; and The Emotions and the Will, 1859, 3rd ed. 1875. HERBERT SPENCER, Principles of Psychology, 1855, 5th ed. 1890. HERBART, Psychologie als Wissenschaft, 2 vols., 1824-1825; and (English trans. by M. K. Smith 1891) Textbook of Psychology,1816.

Works which prepared the way for experimental psychology: LOTZE, Medizinische Psychologie, 1852. G. T. FECHNER, Elemente der Psychophysik, 2 vols., 1860. More extended modern treatises. Of the Herbartian School: W. F. VOLKMANN, Lehrbuch der Psychologie, 2 vols., 4th ed., 1894. M. LAZARUS, Leben der Seele in Monographien, 3 vols., 3rd ed. 1883. Of the Association School (generally with a tendency toward psycho-physical materialism): KUELPE, (English trans. by E. B. Titchener, 1901) Outlines of Psychology, 1893. EBBINGHAUS, Grundzuge der Psychologie, 1st vol. only as yet 1897-1902. ZIEHEN, (English ~trans. by VAN LIEW and BEYER 1899) Introduction to the Study of Physiological Psychology, 6th Ger. ed. 1902. MUNSTERBERG, Grundzuge der Psychologie, 1st vol. only as yet, 1900. Works standing between association psychology and voluntaristic psychology: HOEFFDING, (English trans. by Lowndes, 1891, from the German trans. 1887) Outlines of Psychology, 2nd Danish ed. 1893. W. JERUSALEM, Lehrbuch der empirischen Psychologie, 2nd ed. 1890. Works representing a form of intellectualism related in method to scholasticism: BRENTANO, Psychologie vom empirischen Standpunkte, 1st vol. only, 1874. MEINONG, Psychologischethische Untersuchungen zur Werttheorie, 1894; and Untersuchungen zur Gegenstands theorie und Psychologie, 1904. Works emphasizing the

3. METHODS OF PSYCHOLOGY

1. Since psychology has for its object, not specific contents of experience, but *general experience in its immediate character,* it can make use of no methods except such as the empirical sciences in general employ for the determination, analysis, and causal synthesis of facts. The circumstance, that natural science abstracts from the subject, while psychology does not, can be no ground for modifications in the essential character of the methods employed in the two fields, though it does modify the way in which these methods are applied.

The natural sciences, which may serve as an example for psychology in this respect, since they were developed earlier, make use of *two* chief methods: *experiment and observation. Experiment* is observation connected with an intentional interference on the part of the observer, in the rise and course of the phenomena observed. *Observation,* in its proper sense, is the investigation of phenomena without such interference, just as they are naturally presented to the observer in the continuity of experience. Wherever

independence of psychology and based on an empirical analysis of conscious processes: Lipps, Grundtatsachen des Seelenlebens, 1883; and Leitfaden der Psychologie, 1903. JODL, Lehrbuch der Psychologie, 2nd ed., 1902. The same empirical analysis, and on the basis of this analysis voluntaristic psychology in the sense above described, are presented by the author of this *Outlines of Psychology* in his other works also, namely, Grundzuge der physiologischen Psychologie, 3 vols., 5th ed. 19021903 (English trans. in preparation by E. B. Titchener); and (English trans. by E. B. Creighton and E. B. Titchener, 1894) Lectures on Human and Animal Psychology, 3rd Ger. cd. 1897. Works treating chiefly of the philosophical character of fundamental psychological concepts: UPHUES, Psychologie des Erkennens, 1893. J. REHMKE, Lehrbuch der allgemeinen Psychologie, 1894. NATORP, Einleitung in die Psychologie, 1888. American, English and French works all follow in the path of associationalisrn. Furthermore, they tend for the most part toward psychophysical materialism or toward dualistic spiritualism, less frequently toward voluntarism. From among the numerous American works, the following are to be mentioned: JAMES, Principles of Psychology, 2 vols., 1890. LADD, Psychology Descriptive and Explanatory, 1894. BALDWIN, Handbook of Psychology, 1889. SCRIPTURE, The New Psychology, 1897. TITCHENER, An Outline of Psychology, 1896. French works are as follows: RIBOT'S monographs on various psychological subjects are to be mentioned. (All translated into English: Attention, The Diseases of Memory, The Diseases of the Will, The Diseases of Personality, General Ideas, The Creative Imagination). Also, the works of FOUILLEE, which are related to German voluntarism, but contain at the same time a great deal of metaphysics and are somewhat influenced by the Platonic doctrine of ideas (L'evolutionisme des idees-forces, 1890, and Psychologie des idees-forces, 1893). Works on the history of psychology especially worthy of mention: SIEBECK, Geschichte der Psychologie, Pt. 1st, 18801884, and also articles in the first three vols. of Arch. f. Gesch. d. Phil. (these cover the ancient and medieval periods). LANGE, History of Materialism. DESSOIR, Geschichte der neueren deutschen Psychologie, 2nd ed. 1902 (including as yet only 1st vol.). SOMMER, Grundzuge einer Geschichte der deutschen Psychologie und Aesthetik von Wolf-Baumgarten bis Kant-Schiller, 1892. RIBOT, (English trans. by Baldwin) German Psychology of Today, Fr. ed. 1885, Eng. ed. 1886. W. WUNDT, "Psychologie" in the Festschrift for Kuno Fischer, 1904.

experiment is possible, it is always used in the natural sciences; for under all circumstances, even when the phenomena in themselves present the conditions for sufficiently exact observation, it is an advantage to be able to control at will their rise and progress, or to isolate the various components of a composite phenomenon. Still, even in the natural sciences the two methods have been distinguished according to their spheres of application. It is held that the experimental methods are indispensable for certain problems, while in others the desired end may not infrequently be reached through mere observation. If we neglect a few exceptional cases due to special relations, these two classes of problems correspond to the general division of natural phenomena into *processes* and *objects.*

Experimental interference is required in the exact determination of the course, and in the analysis of the components, of any natural process such as, for example, light-waves or sound-waves, an electric discharge, the formation or disintegration of a chemical compound, and stimulation and metabolism in plants and animals. As a rule, such interference is desirable because exact observation is possible only when the observer can determine the moment at which the process shall commence. It is also indispensable in separating the various components of a complex phenomenon from one another. As a rule, this is possible only through the addition or subtraction of certain conditions, or a quantitative variation of them.

The case is different with *objects* of nature. They are relatively constant; they do not have to be produced at a particular moment, but are always at the observer's disposal and ready for examination. Here, then, experimental investigation is generally necessary only when the production and modification of the objects are to be inquired into. In such a case, they are regarded either as products or components of natural processes and come under the head of processes rather than objects. When, on the contrary, the only question is the actual nature of these objects, without reference to their origin or modification, mere observation is generally enough. Thus, mineralogy, botany, zoology, anatomy, and geography, are pure sciences of observation so long as they are kept free from the physical, chemical, and physiological problems that are, indeed, frequently brought into them, but have to do with processes of nature, not with the objects in themselves.

2. If we apply these considerations to psychology, it is obvious at once, from the very nature of its subject-matter, that exact observation is here possible only in the form of *experimental* observation; and that psychology can never be a *pure* science of observation. The contents of this science are exclusively *processes,* not permanent objects. In order to investigate with exactness the rise and progress of these processes, their composition out of various components, and the interrelations of these components, we must be able first of all to bring about their beginning at will, and purposely to vary the conditions of the same. This is possible here, as in all cases, only through experiment, not through pure introspection. Besides this general reason there is another, peculiar to psychology, that does not apply at all to natural phenomena. In the latter case we purposely abstract from the perceiving subject, and under circumstances, especially when favored by the regularity of the phenomena, as in

astronomy, mere observation may succeed in determining with adequate certainty the objective components of the processes. Psychology, on the contrary, is debarred from this abstraction by its fundamental principles, and the conditions for chance observation can be suitable only when the same objective components of immediate experience are frequently repeated in connection with the same subjective states. It is hardly to be expected, in view of the great complexity of psychical processes, that this will ever be the case. The coincidence is especially improbable since the very *intention to observe,* which is a necessary condition of all observation, modifies essentially the rise and progress of psychical processes. Observation of nature is not disturbed by this intention on the part of the observer, because here we purposely abstract from the state of the subject. The chief problem of psychology, however, is the exact observation of the rise and progress of subjective processes, and it can be readily seen that under such circumstances the intention to observe either essentially modifies the facts to be observed, or completely suppresses them. On the other hand, psychology, by the very way in which psychical processes originate, is led, just as physics and physiology are, to employ the experimental mode of procedure. A sensation arises in us under the most favorable conditions for observation when it is caused by an external sense-stimulus, as, for example, a tone-sensation from an external tone-vibration, or a light-sensation from an external light-impression. The idea of an object is always caused originally by the more or less complicated cooperation of external sense-stimuli. If we wish to study the way in which an idea is formed, we can choose no other method than that of imitating this natural process. In doing this, we have at the same time the great advantage of being able to modify the idea itself by changing at will the combination of the impressions that cooperate to form it, and of thus learning what influence each single condition exercises on the product. Memory-images, it is true, cannot be directly aroused through external sense impressions, but follow them after a longer or shorter interval. Still, it is obvious that their attributes, and especially their relation to the primary ideas through direct impressions, can be most accurately be learned, not by waiting for their chance arrival, but by using such memory-ideas as may be aroused in a systematic, experimental way, through immediately preceding impressions. The same is true of feelings and volitions; they will be presented in the form best adapted to exact investigation when those impressions are purposely produced which experience has shown to be regularly connected with affective and volitional reactions. There is, then, no fundamental psychical process to which experimental methods can not be applied, and therefore none in whose investigation they are not logically required.

3. *Pure observation,* such as is possible in many departments of natural science, is, from the very character of psychic phenomena, impossible in *individual* psychology. Such a possibility would be conceivable only under the condition that there existed permanent psychical objects, independent of our attention, similar to the relatively permanent objects of nature, which remain unchanged by our observation of them. There are, indeed, certain facts at the disposal of psychology, which, although they are not real objects, still have the character of psychical objects inasmuch as they possess these attributes of relative permanence, and independence of the observer. Connected with these characteristics is the further fact that they are unapproachable by means of experiment in the common acceptance of the term. These facts are the *mental products*

that have been developed in the course of history, such as language, mythological ideas, and customs. The origin and development of these products depend in every case on general psychical conditions which may be inferred from their objective attributes. Psychological analysis can, consequently, explain the psychical processes operative in their formation and development. All such mental products of a general character presuppose as a condition the existence of a mental *community* composed of many individuals, though, of course, their deepest sources are the psychical attributes of the individual. Because of this dependence on the community, in particular the social community, this whole department of psychological investigation is designated as *social psychology,* and distinguished from individual, or as it may be called because of its predominating method, *experimental* psychology. In the present stage of the science these two branches of psychology are generally taken up in different treatises; still, they are not so much different departments as different *methods.* So-called social psychology corresponds to the method of pure observation, the objects of observation in this case being the mental products. The necessary connection of these products with social communities, which has given to social psychology its name, is due to the fact that the mental products of the individual are of too variable a character to be the subjects of objective observation. The phenomena gain the necessary degree of constancy only when they become collective.

Thus psychology has, like natural science, *two* exact methods: the experimental method, serving for the analysis of simpler psychical processes, and the observation of general mental products, serving for the investigation of the higher psychical processes and developments.

3a. The introduction of the experimental method into psychology was originally due to the modes of procedure in physiology, especially in the physiology of the sense-organs and the nervous system. For this reason experimental psychology is also commonly called "physiological psychology"; and works treating it under this title regularly contain those supplementary facts from the physiology of the nervous system and the sense-organs, which require special discussion with a view to the interests of psychology, though in themselves they belong to physiology alone. "Physiological psychology" is, accordingly, an intermediate discipline which is, however, as the name indicates, primarily *psychology,* and is, apart from the supplementary physiological facts that it presents, just the same as "experimental psychology" in the sense above defined. The attempt sometimes made, to distinguish psychology proper from physiological psychology, by assigning to the first the psychological interpretation of inner experience, and to the second the derivation of this experience from physiological processes, is to be rejected as inadmissible. There is only one kind of causal explanation in psychology, and that is the derivation of more complex psychical processes from simpler ones. In this method of interpretation physiological elements can be used only as supplementary aids, because of the relation between natural science and psychology as above defined (§ 2, 4). Materialistic psychology denies the existence of psychical causality, and substitutes for this problem the other, of explaining psychical processes by brain-physiology. This tendency, which has been shown (§ 2, 10a) to be epistemologically and psychologically untenable, appears among the representatives of both "pure" and "physiological" psychology.

4. GENERAL SURVEY OF THE SUBJECT

1. The immediate contents of experience which constitute the subject-matter of psychology, are under all circumstances processes of a composite character. Sense-perceptions of external objects, memories of such sense-perceptions, feelings, emotions, and volitional acts, are not only continually united in the most various ways, but each of these processes is itself a more or less composite whole. The idea of an external body, for example, is made up of partial ideas of its parts. A tone may be ever so simple, but we localize it in some direction, thus bringing it into connection with the idea of external space, which is highly composite. A feeling or volition is referred to some sensation that aroused the feeling or to an object willed. In dealing with a complex fact of this kind, scientific investigation has *three* problems to be solved in succession. The *first* is the *analysis* of composite processes; the *second* is the *demonstration of the combinations* into which the elements discovered by analysis enter; the *third* is the *investigation of the laws* that are operative in the formation of such combinations.

2. The *second,* or synthetic, problem is made up of several partial problems. In the first place, the psychical elements unite to form composite *psychical compounds* which are separate and relatively independent of one another in the continual flow of psychical processes. Such compounds are, for example, ideas, whether referred directly to external impressions or objects, or interpreted by us as memories of impressions and objects perceived before. Other examples are composite feelings, emotions, or volitions. Then again, these psychical compounds stand in the most various interconnections with one another. Thus, ideas unite to from larger simultaneous ideational complexes or regular successions, while affective and volitional processes form a variety of combinations with one another and with ideational processes. In this way we have the *interconnection of psychical compounds* as a class of synthetical processes of the *second* degree, consisting of a union between the simpler combinations, or those of elements into psychical compounds. The separate psychical interconnections, in turn, unite to form still more comprehensive combinations, which also show a certain regularity in the arrangement of their components. In this way, combinations of a *third* degree arise which we designate by the general name *psychical developments*. They may be divided into developments of a different scope. Developments of a more limited sort are such as relate to a *single mental trend,* for example, the development of the intellectual functions, of the will, or of the feelings, or of merely one special branch of these functions, such as the aesthetic or moral feelings. From a number of such partial series arises the *total development a psychical personality.* Finally, since animals and in a still higher degree human individuals are in continual interrelation, with like beings, there arise above these individual forms the *general psychical development*s. These various

branches of the study of psychical development are in part the psychological foundations of other sciences, such as the theory of knowledge, aesthetics, and ethics, and are, accordingly, treated more appropriately in connection with these. In part they have become special psychological sciences, such as child-psychology, animal and social Psychology. We shall, therefore, in this treatise discuss only those results from the three last mentioned departments which are of the most importance for general psychology.

3. The solution of the last and most general psychological problem, the ascertation of the *laws of psychical phenomena*, depends upon the investigation of all the combination of different degrees, the combination of elements into compounds, of compounds into interconnections, and of interconnections into developments. And as this investigation is the only thing that can teach us the actual position of psychical processes, so we can discover the attributes of *psychical causality,* which finds its expression in these processes, only from the laws followed by the contents of experience and their components in their various combinations.

We have, accordingly, to consider in the following chapters:

       1) Psychical Elements,

       2) Psychical Compounds,

       3) Interconnection of Psychical Compounds,

       4) Psychical Developments,

       5) Psychical Causality and its Laws

CHAPTER 1

PSYCHICAL ELEMENTS

5. CHIEF FORMS AND GENERAL ATTRIBUTES OF PSYCHICAL ELEMENTS

1. All the contents of psychical experience are of a composite character. It follows, therefore, that *psychical elements,* or the absolutely simple and irreducible components of psychical phenomena, can not be found by analysis alone, but only with the aid of abstraction. This abstraction is rendered possible by the fact that the elements are in reality united in different ways. If the element *a* is connected in one case with the elements *b, c, d* . . ., in another with *b', c', d'* . . ., it is possible to abstract it from all the other elements. because [sic] none of them is always united with it. If, for example, we hear a simple tone of a certain pitch and intensity, it may be located now in this direction, now in that, and may be heard, alternately with various other tones. But since the direction is not constant, or the accompanying tone the same, it is possible to abstract from these variable elements, and we have the single tone as a psychical element.

2. As products of psychical analysis, we have *psychical elements of two kinds,* corresponding to the *two* factors contained in immediate experience (1, 2), the objective contents and the experiencing subject. The elements of the objective contents we call *sensational elements,* or simply *sensations:* such are a tone, or a particular sensation of hot, cold, or light, when we neglect for the moment all the connections of these sensations with others, and all their spacial and temporal relations. The subjective elements, on the other hand, are designated as *affective elements, or simple feelings.* We may mention as examples the feelings accompanying sensations of light, sound, taste, smell, hot, cold, or pain, the feelings aroused by the sight of an agreeable or disagreeable object, and the feelings arising in a state of attention or at the moment of a volitional act. Such simple feelings are in a double sense products of abstraction: each is connected with an ideational element, and is furthermore a component of a psychical process which occurs in time, and during which the feeling itself is continually changing.

3. The actual contents of psychical experience always consist of various combinations of sensational and affective elements, so that the specific character of the simple psychical processes depends for the most part not on the nature of these elements so much as on their union into composite psychical compounds. Thus, the idea of an extended body or of a temporal series of sensations, an emotion, and a volition, are all *specific* forms of psychical experience. But their character as such is as little present in their sensational and affective elements as the chemical properties of a compound body can be defined by recounting the properties of its chemical elements. Specific character and elementary nature of psychical processes are, accordingly, two entirely different concepts. Every psychical element is a specific content of experience, but not every specific content of immediate experience is at the same time a psychical element. Thus, especially spacial and temporal ideas, emotions, and volitional acts, are specific but not elementary processes. Many elements are present only in psychical compounds of a particular kind, but since these compounds regularly contain other elements as well, their special characteristics are to be attributed to the mode of union, rather than to the abstract attributes, of their elements. Thus, we always refer a momentary sound-sensation to a definite point in time. This localization in time, however, is possible only by relating the given sensation to others preceding and following it, so that the special character of the time-idea can not arise from the single sound-sensation thought of as isolated, but only from its union with others. Again, an emotion of anger or a volition contains certain simple feelings that are never present in other psychical compounds, still each of these processes is composite, for it has duration, in the course of which particular feelings follow one another with a certain regularity, and the process itself is not complete without the whole train of these feelings.

4. Sensations and simple feelings exhibit certain common attributes and also certain characteristic differences. They have in common *two determinants,* which we call *quality* and *intensity.* Every simple sensation and every simple feeling has a definite qualitative character that marks it off from all other sensations and feelings; and this quality must always have some degree of intensity. Accordingly, we distinguish the different psychical elements from one another by their qualities, but regard the intensity as the quantitative value which in any concrete case belongs to the given element. Our *designations* of psychical elements are based entirely upon their qualities; thus, we distinguish such sensations as blue, grey, yellow, hot, and cold, or such feelings as grave, cheerful, sad, gloomy, and sorrowful. On the other hand, we always express the differences in the intensity of psychical elements by the same quantitative designations, as weak, strong, medium strong, and very strong. These expressions are in both cases class-concepts which serve for a first superficial arrangement of the elements, and each embraces an unlimitedly large number of concrete elements. Language has developed a relatively complete stock of names for the qualities of simple sensations, especially for colors and tones. Names for the qualities of feelings and for degrees of intensity are far behind. Clearness and obscurity, as also distinctness and indistinctness, are sometimes classed with quality and intensity. But since these attributes, as will appear later (§ 15, 4), always arise from the interconnection of psychical compounds, they can not be regarded as determinants of psychical elements.

5. Made up, at it is, of two determinants, quality and intensity, every psychical element must have a certain *degree of intensity* from which it is possible to pass, by continual gradations, to every other degree of intensity in the same quality. Such gradations can be made in only *two* directions: one we call *increase* in intensity, the other *decrease.* The degrees of intensity of every qualitative element, form in this way a single dimension, in which, from a given point, we may move in two opposite directions, just as from any point in a straight line. This may be expressed in the general statement: *The various intensities of every psychical element form a continuity of one dimension.* The extremities of this continuity we call the *minimal* and *maximal* sensation or feeling, as the case may be.

In contrast with this uniformity in intensifies, the *qualities* have more variable attributes. Every quality may, indeed, be so arranged in a definite continuity that it is possible to pass uninterruptedly from a given point to any other points in the same quality. But the various continuities of different qualities, which we may call *systems of quality,* exhibit differences both in the variety of possible gradations, and in the number of directions of gradation. In these two respects, we may distinguish, on the one hand, *homogeneous and complex,* on the other *one-dimensional, two-dimensional, and many-dimensional* systems of quality. Within a homogeneous system, only such small differences are possible that generally there has never arisen any practical need of distinguishing them by different names. Thus, we distinguish only low quality of pressure, hot, cold, or pain, only one feeling of attention or of activity, although, in intensity, each of these qualities may have many different grades. It is not to be inferred from this fact that in each of these systems there is really only one quality. The truth is that in these cases the number of different qualities is merely very limited; if we were to represent it geometrically, it would probably never reduce entirely to a *single* point. Thus, for example, sensations of pressure from different regions of the skin show, beyond question, small qualitative differences which are great enough to let us distinguish clearly any point of the skin from another at some distance from it. Such differences, however, as arise from contact with a sharp or dull, a rough or smooth body, are not to be reckoned as different qualities. They always depend on a large number of simultaneous sensations, and without the various combinations of these sensations into composite psychical compounds, the impressions mentioned would be impossible.

Complex systems of quality differ from those we have been discussing, in that they embrace a large number of clearly distinguishable elements between which all possible intermediate forms exist. In this class we must include the tonal system and color-system, the systems of smells and tastes, and among the affective systems those which form the subjective complements of these sensational systems, such as the systems of tonal feelings, color-feelings, etc. It is probable also that many systems of feelings belong here, which are objectively connected with composite impressions, but as feelings are simple in character, such are the various feelings of harmony or discord that correspond to the different combinations of tones.

The differences in the *number of dimensions* have been determined with certainty only in the case of certain sensational systems. Thus, the tonal system is one-dimensional. The ordinary color-system, which includes the colors and their transitional qualities to white, is two-dimensional; while the complete system of light-sensations, which includes also the dark color-tones and the transitional qualities to black, is three-dimensional.

6. In the relations discussed thus far, sensational and affective elements in general agree. They differ, on the other hand, in certain essential attributes which are connected with the immediate relations of sensations to objects and of feelings to the subject.

1) When varied in a single dimension, sensational elements exhibit pure *qualitative differences,* which are always in the *same direction* until they reach the possible limits of variation, where they become *maximal differences.* Thus, in the color-system, red and green, blue and yellow, or in the tonal system, the lowest and highest audible tones, are the maximal, and at the ,same time purely qualitative, differences. Every affective element, on the contrary, when continuously varied in the suitable direction of quality, passes gradually into a feeling of *opposite quality.* This is most obvious in the case of the affective elements regularly connected with certain sensational elements, as, for example, tonal feelings or color-feelings. As sensations a high and low-tone are differences that approach more or less the maximal differences of tonal sensation; the corresponding tonal feelings are opposites. In general, then, *sensational qualities are limited by maximal differences, affective qualities by maximal opposites.* Between these opposites is a middle zone, where the feeling is not noticeable it all. It is, however, frequently impossible to demonstrate this indifference-zone, because, while certain simple feelings disappear, other affective qualities remain, or new ones even may arise. The latter case appears especially when the transition of the feeling into the indifference-zone depends on a change in sensations. Thus, in the middle of the musical scale, those feelings disappear which correspond to the high and low tones, but the middle tones have still other, independent affective qualities which do not disappear with these opposites. This is to be explained by the fact that a feeling which corresponds to a certain sensational quality is, as a rule, a component of a complex affective system, in which it belongs at the same time to various dimensions. Thus, the affective quality of a tone of given pitch belongs not only to the dimension of pitch-feelings, but also to that of feelings of intensity and finally to the different dimensions in the clang-qualities of tones may be arraigned. A tone of middle pitch and intensity may, lie in the indifference-zone so far as feelings of pitch and intensity are concerned, and yet have a very marked clang-feeling. The passage of affective elements through the indifference-zone can be directly observed only when care is taken to abstract from other accompanying affective elements. The cases most favorable for this observation are those in which the accompanying elements disappear entirely or almost entirely. Wherever such an indifference-zone appears without complication with other affective elements, we speak of the state as *free from feelings,* and of the sensations and ideas present in such a state, as indifferent.

2) Feelings of specific, and at the same time simple and irreducible, quality appear not only as the subjective complements of simple sensations, but also as the characteristic attendants of composite ideas or even complex ideational processes. Thus, there is a simple tonal feeling which varies with the pitch and intensity of tones, and also a feeling of harmony which, regarded as a feeling, is just as irreducible, but varies with the character of compound clangs. Still other feelings, which may in turn be of the most various kinds, arise from melodious series of clangs. Here, again, each single feeling taken by itself at a given moment, appears as an irreducible unit. Simple feelings are, then, much more various and numerous than simple sensations.

3) The various pure sensations may be arranged in a number of separate system is, between whose elements there is no qualitative relation whatever. Sensations belonging to different systems are called *disparate*. Thus, a tone and a color, a sensation of hot and one of pressure, or, in general, any two sensations between which there are no intermediate qualities, are disparate. According to this criterion, each of the four special senses (smell, taste, hearing, and sight) has a closed, complex sensational system, disparate from the other senses; while the general sense (touch) contains four homogeneous sensational systems (sensations of pressure, hot, cold, and pain). All simple feelings, on the contrary, form a single interconnected manifold, for there is no feeling from which it is not possible to pass to any other through intermediate forms or through indifference-zones. But here too we may distinguish certain systems whose elements are more closely related, as, for example, feelings from colors, tones, harmonies, and rhythms. Still, they are not absolutely closed systems, but there are everywhere relations either of likeness or of opposition to other systems. Thus, such feelings as those from sensations of moderate warmth, from tonal harmony, and from satisfied expectation, however great their qualitative differences may be, are all related in that they belong to the general class of "pleasurable feelings". Still closer relations exist between certain single affective systems, as, for example, between tonal feelings and color-feelings, where deep tones seem to be related to dark colors, and bright colors to high tones. When in such cases a certain relationship is ascribed to the sensations themselves, it is probably due entirely to a confusion of the accompanying feelings with the sensations.

This third distinguishing characteristic shows conclusively that the origin of the feelings is *more unitary* than that of the sensations, which depend on a number of different and in part distinguishable conditions. It is the same distinction that we find in the characterization of the subject, which stands in immediate relation to the feelings, as a unit, in contrast with the plurality of the objects, to which the sensations are related.

6a. It is only in modern psychology that the terms "sensation" and "feeling" have gained the meanings assigned to them in the definitions above given. In older psychological literature they were sometimes used indiscriminatingly, sometimes interchanged. Even yet sensations of touch and those from the internal organs are called feelings by physiologists, and the sense of touch itself is known as the "sense of feeling". This corresponds, it is true, to the original significance of the word, where feeling is the same as touching, still, after the very useful differentiation has once been made, a confusion of the two terms should be avoided. Then again, the word "sensation" is used even by psychologists to mean not only simple, but also composite qualities, such as compound clangs and spacial and temporal ideas. But since we have the entirely adequate word "idea" for such compounds, it is more advantageous to limit the word sensation to psychologically simple sense-qualities. Finally, the term "sensation" has sometimes been restricted so as to mean only those stimulations which come directly from external sense-stimuli. For the psychological attributes of a sensation, however, this circumstance is entirely irrelevant, and therefore such a definition of the term is unjustifiable. The discrimination between sensational and affective elements in any concrete case is very much facilitated by the existence of indifference-zones in the feelings. Then again, from the fact that feelings range between opposites rather than mere differences, it follows that they are much the more variable elements of our immediate experience. This changeable character, which renders it almost impossible to hold an affective state constant in quality and intensity, is the cause of the great difficulties that stand in the way of the exact investigation of feelings.

Sensations are present in all immediate experiences, but feelings may disappear in certain special cases, because of their oscillation through an indifference-zone. Obviously, then, we call, in the case of sensations, abstract from the accompanying feelings, but never vice versa. In this way two false views may easily arise, either that sensations are the *causes* of feelings, or that feelings are a particular species of sensations. The first of these opinions is false because affective elements can never be derived from sensations as such, but only from the attitude of the subject, so that under different subjective conditions the same sensation may be accompanied by different feelings. The second is untenable because the two classes of elements are distinguished, on the one hand by the immediate relation of sensations to objects and of feelings to the subject, and on the other by the fact that the former range between maximal differences, the latter between maximal opposites. Because of the objective and subjective factors belonging to all psychical experience, sensations and feelings are to be looked upon as real and equally essential, though everywhere interrelated, elements of psychical phenomena. In this interrelation the sensational elements appear as the more constant; they alone can be isolated through abstraction, by referring them to external objects. It follows, therefore, of necessity that in investigating the attributes of both, we must start with the sensations. Simple sensations, in the consideration of which we abstract from the accompanying affective elements, are called *pure sensations*. Obviously, we can never speak of "pure feelings" in a similar sense, since simple feelings can never be thought of apart from the accompanying sensations and combinations of sensations. This fact is directly connected with the second distinguishing characteristic mentioned above.

6. PURE SENSATIONS

1. The concept "pure sensation" as shown in § 5 is the product of a twofold abstraction: 1) from the ideas in which the sensation appears, and 2) from the simple feelings with which it is united. We find that pure sensations, defined in this way, form a number of disparate systems of quality; each of these systems, such as that of sensations of pressure, of tone, or of light, is either a homogeneous or a complex continuity (§ 5, 5) from which no transition to any other system can be found.

2. The *rise of sensations,* as physiology teaches us, is regularly dependent on certain physical processes that have their origin partly in the external world surrounding us, partly in certain bodily organs. We designate these processes with a name borrowed from physiology as *sense-stimuli* or *sensation-stimuli.* If the stimulus is a process in the outer world we call it *physical;* if it is a process in our own body we call it *physiological.* Physiological stimuli may be divided, in turn, into *peripheral* and *central*, according as they are processes in the various bodily organs outside of the brain, or processes in the brain itself. In many cases a sensation is attended by all three forms of stimuli. Thus, to illustrate, an external impression of light acts as a physical, stimulus on the eye; in the eye and optic nerve there arises a peripheral physiological stimulation; finally a central physiological stimulation takes place in the corpora quadrigemina and in the occipital regions of the cerebral cortex, where the optic nerve terminates. In many cases the physical stimulus may be wanting, while both forms of physiological stimuli are present; as, when we perceive a flash of light in consequence of a violent ocular movement. In still other cases the central stimulus alone is present; as, when we recall a light impression previously experienced. The central stimulus is, accordingly, the only one that always accompanies sensation. When a peripheral stimulus causes a sensation, it must be connected with a central stimulus, and a physical must be connected with both a peripheral and a central stimulus.

3. The physiological study of development renders it probable that the differentiation of the various sensational systems has been effected in part in the course of general development. The original organ of sense is the outer skin with the sensitive inner organs adjoining it. The organs of taste, smell, hearing, and sight, on the other hand, are later differentiations of it. It may, therefore, be surmised that the sensational systems corresponding to these special sense-organs, have also gradually arisen through differentiation from the sensational systems of the general sense, from sensations of pressure, hot, and cold. It is possible, too, that in lower animals some of the systems now so widely differentiated are even yet more alike. From a physiological standpoint the primordial character of the general sense is also apparent in the fact that it has for the transfer of sense-stimuli to the nerves either very simple organs or none at all. Pressure, temperature, and pain-stimuli can produce sensations at points in the skin where, in spite of the most careful investigation, no special end-organs can be found. There are, indeed, special receiving organs in the regions most sensitive to pressure (touch-

corpuscles, end-bulbs, and corpuscles of Vater), but their structure renders it probable that they merely favor the mechanical transfer of the stimulus to the nerve-endings. Special end-organs for hot, cold, and pain-stimuli have not been found at all.

In the later developed special sense-organs, on the other hand, we find everywhere structures which not only effect the suitable transfer of the stimuli to the sensory nerves, but generally bring about a *physiological transformation* of the stimulation which is indispensable for the rise of the peculiar sensational qualities. But even among the special senses there are differences in this respect.

The receiving organ in the ear, in particular, appears to be of a character different from that of the organs of smell, taste, and sight. In its most primitive forms it consists of a vesicle filled with one or more solid particles (otoliths), and supplied with nerve-bundles distributed in its walls. The particles are set in motion through sound-vibrations, and must cause a rapid succession of weak pressure-stimulations in the fibers of the nerve-bundles. The auditory organ of the higher animals shows an extraordinary complexity, still, in its essential structure it recalls this primitive type. In the cochlea of man and the higher animals the auditory nerve passes at first through the axis, which is pierced by a large number of fine canals, and then emerges through the pores which open into the cavity of the cochlea. Here the branches are distributed on a tightly stretched membrane, which extends through the spiral windings of the cochlea and is weighted with special rigid arches (arches of Corti). This membrane - the basilar membrane, as it is called - must, according to the laws of acoustics, be thrown into sympathetic vibrations whenever sound-waves strike the ear. It seems, therefore, to play the same part here as the otoliths do in the lower forms of the auditory organ. At the same time one other change has taken place which accounts for the enormous differentiation of the sensational system. The basilar membrane has a different breadth in its different parts, for it grows continually wider from the base to the apex of the cochlea. In this way it acts like a system of stretched chords of different lengths. And just as in such a system, other conditions remaining the same, the longer chords are tuned to lower and the shorter to higher tones, so we may assume the same to be true for the different parts of the basilar membrane. We may surmise that the simplest auditory organs with their otoliths have a homogeneous sensational system, analogous perhaps to our systems of sensations of pressure. The special development of the organ as seen in the cochlea of higher animals explains the evolution of an extraordinarily complex sensational system from this originally homogeneous system. Still, the structure remains similar in this respect, that it seems adapted, in the latter case as in the former, to the best possible *transfer* of the physical stimulus to the sensory nerve rather than to any transformation of the stimulus. This view agrees with the observed fact that, just as sensations of pressure may be perceived, on regions of the skin not supplied with special receiving organs, so, in the case of certain animals, such as birds, where the conditions are specially favorable for their transmission, sound-vibrations are transferred to the auditory nerve and sensed even after the removal of the whole auditory organ with its special receiving structure.

With *smell, taste,* and *sight* the case is essentially different. Organs are present which render direct action of the stimuli on the sensory nerves impossible. The external stimuli are here received through special organs and modified before they excite the nerves. These organs are specially metamorphosed epithelial cells with one end exposed to the stimulus and the other passing into a nerve fiber. Everything goes to show that the receiving organs here are not merely for the transfer of the stimuli, but rather for their *transformation*. In the three cases under discussion it is probable that the transformation is a *chemical* process. In smell and taste we have external chemical agencies, in sight we have light as the causes of chemical disintegrations in the sensory cells; these processes in the cells then serve as the real stimuli.

These three senses may, as *chemical* senses, be distinguished from the *mechanical* senses of pressure and sound. It is impossible to say with any degree of certainty, to which of these two classes sensations of cold and hot belong. One indication of the direct relation between stimuli and sensation in mechanical senses, as contrasted with the indirect relation in chemical senses, is that in the first case the sensation lasts only a very little longer than the external stimulus, while in the latter case it persists very much longer. Thus, in a quick succession of pressures and more especially of sounds, it is possible to distinguish clearly the single stimuli from one another; lights, tastes, and smells, on the other hand, run together at a very moderate rate of succession.

4. Since peripheral and central stimuli are regular physical concomitants of elementary sensational processes, the attempt to determine the relation between stimuli and sensations is very natural. In attempting to solve this problem, physiology generally considers sensations as the result of physiological stimuli, but assumes at the same time that in this case any proper explanation of the effect from its cause is impossible, and that all that can be undertaken is to determine the constancy of the relations between particular stimuli, and the resulting sensations. Now, it is found in many cases that different stimuli acting on the same end-organ produce the same sensations; thus, for example, mechanical and electrical stimulations of the eye produce light sensations. This result was generalized in the principle, that every receiving element of a sense-organ and every simple sensory nerve-fiber together with its central terminus, is capable of only a single sensation of fixed quality; that the various qualities of sensation are, therefore, due to the various physiological elements with different specific energies.

This principle, generally called the "law of specific energy of nerves", is untenable for *three* reasons, even if we neglect for the moment the fact that it simply refers the causes of the various differences in sensations to a *qualtalitas occutlta* of sensory and nervous elements.

1) It is contradictory to the physiological doctrine of the development of the senses. If, as we must assume according to this doctrine, the complex sensational systems are derived from systems originally simpler and more homogeneous, the physiological sensory elements must have undergone a change also. This, however, is possible only under the condition that organs may be modified by the stimuli which act upon them. That is to say, the sensory elements determine the qualities of sensations only

31

secondarily, as a result of the properties which they acquire through the processes of stimulation aroused in them. If, then, these sensory elements have undergone, in the course of time, radical changes due to the nature of the stimuli acting upon them, such changes could have been possible only under the condition that the physiological stimulations in the sensory elements varied to some extent with the quality of the stimulus.

2) The principle of specific energy is contradictory to the fact that in many senses the number of different sensory elements does not correspond at all to that of different sensational qualities. Thus, from a single point in the retina we can receive all possible sensations of brightness and color; in the organs of smell and taste we find no clearly distinguishable forms of the sensory elements, while even a limited area of their sensory surfaces can receive a variety of sensations, which, especially in the case of the olfactory organ, is very large. Where we have every reason to assume that qualitatively different sensations actually do arise in different sensory elements, as in the case of the auditory organ, the structure of the organ goes to show that this difference is not due to any attribute of the nerve-fibers or of other sensory elements, but that it comes originally from the way in which they are arranged. Different fibers of the auditory nerve will, of course, be stimulated by different tone-vibrations, because the different parts of the basilar membrane are tuned to different tones; but this is not due to some original and inexplicable attribute of the single auditory nerve-fibers, but to the way in which they are connected with the end-organ.

3) Finally, the sensory nerves and central elements can have no original specific energy, because the peripheral sense-organ must be exposed to the adequate stimuli for a sufficient interval, or must at least have been so exposed at some previous period, before the corresponding sensations can arise through their stimulation. Persons congenitally blind and deaf do not have any sensations of light or tone whatever, so far as we know, even when the sensory nerves and centers were originally present.

Everything goes to show that the differences in the qualities of sensations are conditioned by the differences in the *processes of stimulation* that arise in the sense-organs. These processes are dependent, primarily on the character of the *physical* stimuli, and only secondarily on the peculiarities of the receiving organ, which are due to its adaptation to these stimuli. As a result of this adaptation, however, it may happen that even when some stimulus other than that which has effected the original adaptation of the sensory elements, that is, when an inadequate stimulus acts, the sensation corresponding to the adequate stimulus may arise. Still, this does not hold for all stimuli or for all sensory elements. Thus, hot and cold stimulations can not cause cutaneous sensations of pressure or sensations in the special sense-organs; chemical and electrical stimuli produce sensations of light only when they act upon the retina, not when they act on the optic nerve; and, finally, these general stimuli can not arouse sensations of smell or taste. When an electric current causes chemical disintegration, it may, indeed, arouse such sensations, but it is through the adequate chemical stimuli produced.

5. From the very nature of the case, it is impossible to explain the character of sensations from the character of physical and physiological stimuli. Stimuli and sensations can not be compared with one another at all; the first belong to the mediate experience of the natural sciences, the second to the immediate experience of psychology. An interrelation between sensations and *physiological* stimuli must necessarily exist, however, in the sense that different kinds of stimulation always correspond to different sensations. This *principle of the parallelism of changes in sensation and* in *physiological stimulation* is an important supplementary principle in both the psychological and physiological doctrines of sensation. In the first case it is used in producing definite changes in the sensation by means of intentional variation of the stimulus; in the second it is used in inferring the identity or non-identity of physiological stimulations from the identity or non-identity of the sensations. Furthermore, the same principle is the basis of our practical life and of our theoretical knowledge of the external world.

A. SENSATIONS OF THE GENERAL SENSE

6. The definition of the "general sense" includes two factors. In point of time, the general sense is that which precedes all others and therefore belongs to *all* beings, endowed with mind. In its spacial attributes, the general sense is distinguished from the particular senses in having the most extensive sensory surface exposed to stimuli. It includes not only the whole external skin and the adjoining areas of the mucous membrane, but a large number of internal organs supplied with sensory nerves, such as joints, muscles, tendons, and bones, which are accessible to stimuli either always, or at certain times, under special conditions, as is the case with bones. The general sense includes four specific, distinct sensational systems: sensations of pressure, hot, cold, and pain. Not infrequently a single stimulus arouses more than one, of these sensations. The sensation is then immediately recognized as made up of a mixture of components from the different systems; for example, from sensations of pressure and pain, or from sensations of hot and pain. In a similar manner as a result of the extension of the sense-organ, we may often have mixtures of the various qualities of one and the same system, for example, qualitatively different sensations of pressure, when an extended region of the skin is touched.

The four systems of the general sense are all *homogenous* systems (§ 5, 5). This shows that the sense is genetically earlier than the others, whose systems are all *complex*. The sensations of pressure from the external skin, and those due to the tensions and movements of the muscles, joints, and tendons, are generally grouped together under the name *touch-sensations,* and distinguished from the *common sensations,* which include sensations of hot, cold, and pain, and those sensations of pressure that sometimes arise in the other internal organs. This distinction, however, has its source in the relation of the sensations to ideas and concomitant feelings, and has nothing to do with the qualities of the sensations in themselves.

7. The ability of the different parts of the general sense-organ to receive stimulations and give rise to sensations, can be tested with adequate exactness only on the external skin. The only facts that can be determined in regard to the internal parts, are that the joints are in a high degree sensitive to pressures, while the muscles and tendons are much less so, and that sensations of hot, cold, and pain in the internal organs are exceptional, and noticeable only under abnormal conditions. On the other hand, there is no point of the external skin and of the immediately adjoining parts of the mucous membrane, which is not sensitive at once to stimulations of pressure, hot, cold, and pain. The *degree* of sensitivity may, indeed, vary at different points, in such a way that the points most sensitive to pressure, to hot, and to cold, do not, in generally, coincide. Sensitivity to pain is everywhere about the same, varying at most in such a way that in some places the pain-stimulus acts on the surface, and in others not until it has penetrated deeper. On the other hand, certain approximately punctiform cutaneous regions appear to be most favorable for stimulations of pressure, hot, and cold. These points are called respectively, pressure-spots, hot-spots, and cold-spots. They are distributed in different parts of the skin in varying numbers. Spots of different modality never coincide; still, temperature-spots always receive sensations of pressure and pain as well; and a pointed hot stimulus applied to a cold spot, always causes a sensation of hot, while hot-spots do not seem to be stimulated by pointed cold stimuli. Furthermore, hot-spots and cold-spots react with their adequate sensations to properly applied mechanical and electrical stimuli.

8. Of the four qualities mentioned sensations of pressure and pain form closed systems which show no relations either to each other or to the two systems of temperature-sensations. These last two, on the other hand, stand in the relation of *opposites;* we apprehend hot and cold not merely as different, but as *contrasted* sensations. It is, however, very probable that this is not due to the original nature of the sensations, but partly to the conditions of their rise, and partly to the accompanying feelings. For, while the other qualities may be united without limitation to form mixed sensations ~ as, for example, pressure and hot, pressure and pain, cold and pain ~ hot and cold exclude each other because, under the conditions of their rise, the only possibilities for a given cutaneous region are a sensation of hot or one of cold, or else an absence of both. When one of these sensations passes continuously into the other, the change regularly takes place in such a way that either the sensation of hot gradually disappears and a continually increasing sensation of cold arises, or vice versa the sensation of cold disappears and that of hot gradually arises. Then, too, elementary feelings of opposite character are connected with hot and cold, the point where both sensations are absent corresponding to their indifferent zone.

In still another respect the two systems of temperature-sensations are peculiar. They are to a great extent dependent on the varying conditions under which the stimuli act upon the sense-organ. A considerable increase above the temperature of the skin is perceived as hot, while a considerable decrease below the same is perceived as cold, but the temperature of the skin itself, which is the indifference zone between the two, can adapt itself rapidly to the existing, external temperature within fairly wide limits. The fact that in this respect too, both systems are alike, favors the view that they are interconnected and also antagonistic.

B. SENSATIONS OF SOUND

9. We possess two independent systems of simple auditory sensations, which are generally, however, connected as a result of the mixture of the two kinds of impressions. They are the *homogeneous* system of simple noise-sensations and the *complex* system of simple tone-sensations.

Simple noise-sensations can be produced only under conditions that exclude the simultaneous rise of tonal sensations, as when air-vibrations are produced whose rate is either too rapid or too slow for tone-sensations to arise, or when the sound-waves act upon the ear for too short a period. Simple sensations of noise, thus produced, may vary in intensity and duration, but apart from these differences they are qualitatively alike. It is possible that small qualitative differences also exist among them, due to the conditions of their rise, but such differences are too small to be marked by distinguishing names. The noises commonly so called are compound ideas made up of such simple noise-sensations and of a great many irregular tonal sensations (cf. § 9, 7). The homogeneous system of simple noise-sensations is probably the first to develop. The auditory vesicles of the lower animals, with their simple otoliths, could hardly produce anything but these. In the case of man and the higher animals it may be surmised that the structures found in the vestibule of the labyrinth receive only homogeneous stimulations, corresponding to simple sensations of noise. Finally, experiments with animals deprived of their labyrinths, make it probable that even direct stimulations of the auditory nerve can produce such sensations. In the embryonic development of the higher animals, the cochlea develops from an original, vestibular vesicle, which corresponds exactly to a primitive auditory organ. We are, therefore, justified in supposing that the complex system of tonal sensations is a product of the differentiation of the homogeneous system of simple noise-sensations, but that in every case where this development, has taken place, the simple system has remained along with the higher.

10. The system of *simple tone-sensation* is a continuity of one dimension. We call the quality of the single simple tones *pitch.* The one-dimensional character of the system finds expression in the fact that, starting with a given pitch, we can vary the quality only in *two* opposite directions: one we call *raising* the pitch, the other *lowering* it. In actual experience simple sensations of tone are never presented alone, but always united with other tonal sensations and with accompanying simple sensations of noise. But since, according to the scheme given above (§ 5, 1), these concomitant elements can be varied indefinitely, and since in many cases they are relatively weak in comparison with one of the tones, the abstraction of simple tones was early reached through the practical use of tonal sensations in the art of music. The names *c, c#, d#,* and d stand for simple tones, though the clangs of musical instruments or of the human voice by means of which we produce these different pitches, are always accompanied by other, weaker tones and often, too, by noises. But since the conditions for the rise of such concomitant

tones can be so varied that they become very weak, it has been possible to produce really simple tones of nearly perfect purity. The simplest means of doing this is by using a tuning-fork, and a resonator tuned to its fundamental tone. Since the resonator increases the intensity of the fundamental only, the other, accompanying tones are so weak when the fork sounds, that the sensation is generally apprehended as simple and irreducible. If the sound-vibrations corresponding to such a tonal sensation are examined, they will be found to correspond to the simplest possible form of vibration, the pendulum-oscillation, so called because the vibrations of the atmospheric particles follow the same laws as a pendulum oscillating in a very small amplitude.[2] That these relatively simple sound-vibrations correspond to sensations of simple tones, and that we can even distinguish the separate tones in compounds, can be explained, on the basis of the physical laws of sympathetic vibration, from the structure of the organs in the cochlea. The basilar membrane in the cochlea is in its different parts tuned to tones of different pitch, so that when a simple oscillatory sound-vibration strikes the ear, only the part tuned to that particular pitch will vibrate in sympathy. If the same rate of oscillation comes in a compound sound-vibration, again only the part tuned to it will be affected by it, while the other components of the wave will set in vibration other sections of the membrane, which correspond in the same way to their pitch.

11. The system of tonal sensations shows its character as a *continuous* series in the fact that it is always possible to pass from a given pitch to any other through continuous changes in sensation. Music has selected at option from this continuity single sensations separated by considerable intervals, thus substituting a *tonal scale* for the *tonal line.* This selection, however, is based on the relations of tonal sensations themselves. We shall return to the discussion of these relations later, in taking up the ideational compounds arising from these sensations (§ 9). The natural tonal line has two extremities, which are conditioned by the physiological capacity of the ear for receiving sounds. These extremities are the *lowest* and *highest* tones; the former corresponds to 8-10 double vibrations per second, the latter to 40,000-50,000.

C. SENSATIONS OF SMELL AND TASTE

12. *Sensations of smell* form a complex system whose arrangement is still unknown. All we know is that there is a very great number of olfactory qualities, between which there are all possible transitional forms. There can, then, be no doubt that the system is a continuity of many dimensions.

12a. Olfactory qualities may be grouped in certain *classes,* each of which contains *classes* sensations which are more or less related. This fact may be regarded as an indication of how these sensations may perhaps be reduced to a small number of principal qualities. Such classes are for examples, sensations like those from ether, balsam, musk, benzine, those known as aromatic, etc.

---

[2] Pendulum-oscillations may be represented by a *sine-curve,* because the distance from the position of rest is always proportional to the sine of the time required to swing to the point in question.

It has been observed in a few cases that certain olfactory sensations which come from definite substances, can also be produced by mixing others. But these observations are still insufficient to reduce the great number of simple, qualities contained in each of the classes mentioned, to a limited number of primary qualities and their mixtures. Finally, it has been observed that many odors neutralize each other, so far as the sensation is concerned, when they are mixed in the proper intensities. This is true not only of substances that neutralize each other chemically, as acetic acid and ammonia, but also of others, such as caoutchoue and wax or tolu-balsam, which do not act on each other chemically outside of the olfactory cells. Since this neutralization takes place when the two stimuli act on entirely different olfactory surfaces, one on the right and the other on the left mucous membrane of the nose, it is probable that we are dealing, not with phenomena analogous to those exhibited by complementary colors (22), but with a reciprocal central inhibition of sensations. Another observed fact tells against the notion that they are complementary. One and the same olfactory quality can neutralize several entirely different qualities, sometimes even those which in turn neutralize one another, while among colors it is always only *two* fixed qualities that are complementary.

13. *Sensations of taste* have been somewhat more thoroughly investigated, and we can here distinguish *four*, distinct *primary qualities*. Between these there, are all possible transitional tastes, which are to be regarded is mixed sensations. The primary qualities are *sour, sweet, bitter,* and *saline*. Besides these, alkaline and metallic are sometimes regarded as independent qualities. But alkaline qualities show an unmistakable relationship with saline, and metallic with sour, so that both are probably mixed sensations, (alkaline made up perhaps of saline and sweet, metallic of sour and saline). Sweet and saline are opposite qualities. When these two sensations are united in proper intensifies, the result is a mixed sensation (commonly known as "insipid"), even though the stimuli that here reciprocally neutralize each other do not enter into a chemical combination. The system of taste-sensations is, accordingly, in all probability to be regarded as *a two-dimensional* continuity, which may be geometrically represented by a circular surface on whose circumference, the four primary, and their intermediate, qualities are arranged, while the neutral mixed sensation is in the middle, and the other transitional taste-qualities on the surface, between this middle point and the saturated qualities on the circumference.

13a. In these attributes of taste-qualities we seem to have the fundamental type of a chemical sense. In this respect taste is perhaps the antecedent of sight. The obvious interconnection with the chemical nature of the stimulation, makes it probable even here that the reciprocal neutralization of certain sensations, with which the two-dimensional character of the sensational system is perhaps connected, depends, not on the sensations in themselves, but on the relations between the *physiological* stimulations, just as in the case of sensations of hot and cold. It is well known that very commonly the chemical effect of certain substances can be neutralized through the action of certain other substances. Now, we do not know what the chemical changes are that are produced by the gustatory stimuli in the taste-cells. But from the neutralization of sensations of sweet and saline we way conclude, in accordance with the principle of the parallelism of changes in sensation and in stimuli, that the chemical reactions which sweet and saline substances produce in the sensory cells, also counteract each other. The same would hold for their sensations for which similar relations could be demonstrated. In regard to the physiological conditions for gustatory stimulations, we can draw only this one conclusion from the facts mentioned, namely, that the chemical processes of stimulation corresponding to the sensations which neutralize each other in this way, probably take place in the same cells. Of course, the possibility is not excluded that several different processes liable to neutralization through opposite reactions, could arise in the same cells. The known anatomical facts and the experiments of physiology in stimulating single papillae separately, give lie certain conclusions in this matter. Whether we are here dealing with phenomena that are really analogous to those exhibited by complementary colors is still a question.

D. SENSATIONS OF LIGHT

14. The system of light-sensations is made up of *two* partial systems: that of *sensation of achromatic light* and that of *sensations of chromatic light*. Between the qualities in these two, all possible transitional forms exist.

Sensations of achromatic light, when considered alone, form a complex system of one dimension, which extends, like the tonal line, between two limiting qualities. The sensations in the neighborhood of one of these limits we call *black;* in the neighborhood of the other white, while between the two we insert grey in its different shades (dark grey, grey, and light grey). This one-dimensional system of achromatic sensations differs from that of tones in being *at once a system of quality and of intensity* for every qualitative change in the direction from black to white is seen at the same time as an increase in intensity, an every qualitative change in the direction from white to black is seen as a decrease intensity. Each point in the series, which thus has a definite quality and intensity, is called a degree of *brightness* of the achromatic sensations. The whole system may, accordingly, be designated as the *sensations of pure brightness.* The use of the word "pure" indicates the absence of all sensations of color. The system of pure brightness is absolutely one-dimensional for, both the variations in quality and those in intensity belong to one and the same dimension. It differs essentially, in this respect, from the tonal line, in which each point is merely a degree of quality, and has also a whole series of gradations in intensity. Simple tone-sensations thus form a two-dimensional continuity so soon as we take into account both determinants, quality and intensity, while the system of pure brightness is always *one-dimensional,* even when we attend to both determinants. The whole system may, therefore, be regarded as a continuous series of *grades of brightness,* in which the lower grades are designated black so far as quality is concerned, and weak in point of intensity, while the higher grades are called white and strong.

15. *Sensations of color* also form a one-dimensional system when their qualities alone are taken into account. Unlike the system of sensations of pure brightness, this system returns upon itself from whatever point we start, for at first, after leaving a given quality, we pass gradually to a quality that shows the greatest difference, and going still

38

further we find that the qualitative differences become smaller again, until finally we reach the starting point once more. The color-spectrum obtained by refracting sunlight through a prism, or that seen in the rainbow, shows this characteristic, though not completely. If in these cases we start from the red end of the spectrum, we come first to orange, then to yellow, yellow-green, green-blue, blue, indigo-blue, and finally to violet which is more like red than any of the other colors except orange, which lies next to red. The line of colors in the spectrum does not return quite to its starting-point, because it does not contain all of colors that we have in sensation. Purple-red shades, which can be obtained by the objective mixture of red and violet rays, are wanting in the spectrum. Only when we fill out the spectral series with them, is the system of actual color-sensations complete, and then the system is a closed circle. This characteristic is not to be attributed to the circumstance that the spectrum actually presents for our observation a series returning nearly to its beginning. The same order of sensations can be found by arranging according to their subjective relationship, colored objects presented in any irregular order. Even children who have never observed attentively a solar spectrum or a rainbow, and can, therefore, begin the series with any other color just as well as with red, always arrange them in the same order.

The system of pure colors is, then, to be defined as one-dimensional. It does not extend in a straight line, however, but *returns upon itself.* Its simplest geometrical representation would be a *circle.* From a given point in this system we pass, when the sensation is gradually varied, first to sensations, then to those most markedly different, and finally to others similar to the first quality, but in the opposite direction. Every color must, accordingly, be related to *one* other particular color as a *maximum of difference* in sensation. This color may be called the *opposite color,* and in the representation of the color-system by a circle, two opposite colors are to be placed at the two extremities of the same diameter. Thus, for example, purple-red and green, yellow and blue, light green and violet, are opposite colors, that is, colors which exhibit the greatest qualitative differences.

The quality determined by the position of a sensation in the color-system, in distinction to other qualitative determinations, is called *color-tone,* a figurative name borrowed from tonal sensations. In this sense the simple names of colors; such as red, orange, yellow, etc., denote merely color tones. The color-circle is a representation of the system of color-tones abstracted from all the other attributes belonging to the sensations. In reality, every color-sensation has *two* other attributes, one we call its *saturation,* the other its *brightness.* Saturation is peculiar to chromatic sensations, while brightness belongs to achromatic sensations as well.

16. By *saturation* we mean the attribute of color-sensations by virtue of which they appear in all possible stages of transition to sensations of pure brightness, so that a continuous passage is possible from every color to any point in the series of whites, grays, and blacks. The term "saturation" is borrowed from the common method of producing these transitional colors objectively, that is, by the more or less intense saturation of some colorless soluble with color-pigment. A color may be ever so saturated, yet it is possible to think of a still greater saturation of the same color-tone,

and, on the other hand, pure brightness always denotes the end of the series of diminishing grades of saturation for any color whatever. A *degree of saturation* may, therefore, be thought of as an attribute of all color-sensations, and, at the same time, as the attribute by which the system of color-sensations is directly united with that of sensations of pure brightness. If, now, we represent some particular sensation of white, grey, or black by the central point of the color-circle, all the grades of saturation that can arise as transitional stages from any particular color to this particular sensation of pure brightness, will obviously be represented by that radius of the circle which connects the center with the color in question. If the shades of saturation corresponding to the continuous transitional stage, from all the colors to a particular sensation of pure brightness are thus geometrically represented, we have the system of saturation-grades as a *circular surface* whose circumference is a system of simple color-tones, and whose center is the sensation of pure brightness, corresponding to the absence of all saturation. For the formation of such a system of saturation-grades any point whatever in the series of sensations of pure brightness may be taken, so long as the condition is fulfilled that white is not too bright or the black too dark, for in such differences in both saturation and color disappear. Systems of saturation which are arranged about *different* points in the series of pure brightness, always have different grades of brightness. A *pure* system of saturation, accordingly, call be made for only *one* particular grade of brightness at a time, that is, for only one point in the series of sensations of pure brightness. When such systems are made for all possible points, the system of saturation will be supplemented by that *of grades of brightness.*

17. *Brightness* is just as an attribute of color-sensation as it is of achromatic sensations, and is in this case, too, at once a quality and degree of intensity. Starting from a given grade, if the brightness increases, every color approaches white, in quality, while at the same time the intensity increases; if the brightness decreases, the colors approach black in quality, and the intensity diminishes. The grades of brightness for any single color thus form a system of intensive qualities, analogous to that of pure brightness, only in place of the achromatic gradations between white and black, we have the corresponding grades of saturation. From the point of greatest saturation there are *two* opposite for variation in saturation: one *positive,* towards white, accompanied by an increase in the intensity of the sensation, and the other *negative,* towards black, with a corresponding decrease in intensity. As limits for these two directions we have, on the one hand, the pure sensation white, on the other, the pure sensation black; the first is at the same time the maximum, the second the minimum of intensity. White and black are in this way opposite extremities of the system of sensations of pure brightness, and also of the system of color-sensations arranged according to grades of brightness. It follows obviously that there is a certain medium brightness for every color, at which its saturation is greatest. From this point, the saturation diminishes in the positive direction when the brightness increases, and in the negative direction when the brightness decreases. The grade of brightness most favorable for the saturation is not the same for all colors, but varies from red to blue, in such a way that it is most intense for red and least intense for blue. This accounts for the familiar phenomenon that in twilight, when the degree of brightness is small, the blue color-tones - of paintings, for example - are still clearly visible, while the red color-tones appear black.

18. If we neglect the somewhat different position of the maximal saturation of the various colors in the line of brightness, the relation that exists between *sensations of chromatic brightness and those of pure, or achromatic, brightness,* by virtue of the gradual transition of colors into white on the one hand, and into black on the other, may be represented in the simplest manner as follows. First, we may represent the system of pure color-tones, that is, of the colors at their maximal saturation, by a circle, as above. Then we may draw through the center of this circle, perpendicular to its plane, the straight line of pure brightness, in such a way that where it cuts the plane of the circular surface, it represents the sensation of pure brightness corresponding to the minimum of saturation for the colors with which we started. In like manner, the other color-circles for increasing and decreasing grades of brightness, may be arranged perpendicularly along this line, above and below the circle of greatest saturation. But the decreasing saturation of the colors in these latter circles must be expressed in the shortening of their radii; just as in the first circle, the shorter the distance from the center, the less the saturation. These radii grow continually shorter, until finally, at the two extremities of the line, the circles disappear entirely. This corresponds to the fact that for every color the maximum of brightness corresponds to the sensation white, while its minimum corresponds to black.[3]

19. The whole system of *sensations of* chromatic *brightness* may, accordingly, be most simply represented by a *spherical surface* whose equator represents the system of pure color-tones, or colors of greatest saturation, while the two poles correspond to white and black, the extremities of the sensations of chromatic brightness. Of course, any other geometrical figure with similar attributes, as, for example, two cones with a common base and apexes pointing in different directions, would serve the same purpose. The only thing essential for the representation, is the gradual transition to white and black, and the corresponding decrease in the variety of the color-tones, which finds its expression in the continual decrease in the length of the radii of the color-circles. Now, as above shown, the system of saturations corresponding to a particular sensation of pure brightness, may be represented by a circular surface which contains all the sensations of light belonging to one grade of brightness. When we unite grades of saturation and brightness to a single system, the *total system of light sensations* may be represented by a *solid sphere*. The equator is the system of pure color-tones; the polar axis is the system of pure brightnesses; the surface represents the system of chromatic brightnesses, and, finally, every circular plane perpendicular to the polar axis, corresponds to a system of saturations of equal brightness. This representation by means of a sphere is indeed arbitrary, in the sense that any other solid figure with analogous attributes may be chosen in its place; still, it presents to view the psychological fact that the *total system of light sensations is a closed continuity of three dimensions.* The *three-dimensional* character of the system arises from the fact that every concrete sensation of light has three determinants: color-tone, saturation, and brightness. Pure, or achromatic, brightness and pure, or saturated, colors are to be

---

[3] It must be observed, however, that the actual coincidence of these sensations can be empirically proved only for the minimum of brightness. Grades of brightness which approach the maximum are so injurious to the eye that the general demonstration of the approach to white must be accepted as sufficient.

regarded as the two extreme cases in the series of saturations. The *closed* form of the system comes from the circular character of the color-line, on the one hand, and, on the other, from the termination of the system of chromatic brightness in the extremes of pure brightness. A special characteristic of the system is that only the changes in the *two* dimensions, or those of color-tones and saturations, are pure variations in quality, while every movement in the *third* dimension, or that of brightness, is at once a modification of both quality and intensity. As a consequence of this circumstance, the whole three-dimensional system is required to represent the qualities of light-sensations, but it includes also the intensities of these sensations.

20. Certain *principal senses* are prominent in this system, because we use them as points of reference for the arrangement of all the others. These are, *white* and *black*, in the achromatic series, and the four principal colors, *red, yellow, green,* and *blue*, in the chromatic. Only these six sensations have clearly distinguished names in the early development of language. All other sensations are then named either with reference to these or with modifications of the names themselves. Thus, we regard grey as a stage in the achromatic series lying between white and black, We designate the different grades of saturation according to their brightness, as whitish or blackish, light or dark color-tones; an we generally choose compound names for the colors between the four principal ones, as, for example, purple-red, orange-yellow, yellow-green, etc. These all show their relatively late origin by their very composition.

20a. From the early origin of the names for the six qualities mentioned, the conclusion has been drawn that they are *fundamental* qualities of vision, and that the others are compounded from them. Grey is declared to be a mixture of black and white, violet and purple-red to be mixtures of blue and red, etc. Psychologically there is no justification for calling any light-sensations compound in comparison with others. Grey is a simple sensation just as much as white or black; such colors as orange and purple-red are just as much simple colors as red and yellow; and any grade of saturation which we have placed in the system between a pure color and white, is by no means, for that reason, a compound sensation. The closed, continuous character of the system makes it necessary for language to pick out certain especially marked differences in reference to which all other sensations are then arranged, for the simple reason that it is impossible to have an unlimited number of names. It is most natural that white and black should be chosen as such points of reference for the achromatic series, since they designate the greatest differences. When once these two are given, however, all other achromatic sensations will be considered as transitional sensations between them, since the extreme differences are connected by a series of all possible grades of brightness. The case of color-sensations is similar; only here, on account of the circular form of the color-line, it is impossible to choose directly two absolutely greatest differences. Other motives besides the necessary qualitative difference, are decisive in the choice of the principal colors. We may regard as such motives, the frequency and affective intensity of certain light-impressions due to the natural conditions of human existence. The red color of blood, the green of vegetation, the blue of the sky, and the yellow of the heavenly bodies in contrast with the blue of the sky, and the yellow heavenly bodies may well have furnished the earliest occasions for the choice of certain colors as those to receive names. Language generally names the sensation from the object that produced it, not the object from the sensation. In this case too, when certain principal qualities were once determined, all others must, on account of the continuity of the series of sensations, seem to be intermediate color-tones. The difference between principal colors and transitional colors is, therefore, very probably due entirely to external conditions. If these conditions had been other, red might have been regarded as a transitional color between purple and orange, just as orange is now placed between red and yellows.[4]

21. The attributes of the system of light-sensations above described, are so peculiar as to lead us to expect *a priori* that the relation between these psychological attributes and the objective processes of stimulation, is essentially different from that in the cases of the sensational systems discussed before, especially those of the general and auditory senses. Most striking, in this respect, is the difference between the system in question and that of tones. In the latter case, the principle of parallelism between sensation and stimulus, holds not only for the physiological processes of. stimulation, but to a great extent for the physical processes as well. A simple sensation corresponds to a simple form of sound-vibration, and a plurality of simple sensations to compound form. Furthermore, the intensity of the sensation varies in proportion to the amplitude of the vibrations, and its quality with their form, so that in both directions the subjective difference between sensations increases with the growing difference between the objective physical stimuli. The relation in the case of light-sensations is entirely different. Like objective sound, objective light also consists of vibrations in some medium. To be sure, the actual form of these vibrations is still a question, but from physical experiments on the phenomena of interference we know that the consist of very short and rapid waves. Those seen as light vary in wave-length from 688 to 393 millionths of a millimeter, and in rate from 450 to 790 billion vibrations per second. In this case, too, simple sensations correspond to simple vibrations, that is, vibrations of like wave-length; and the quality of the sensation varies continuously with the rate: red corresponds to the longest and slowest wives, and violet to the shortest and most rapid, while the other color-tones form a continuous series between these, varying with the changes in wave-length. Even here, however, an essential difference appears, for the colors red and violet, which are the most different in wave-length, are more similar in sensation than those which lie between.[5]

There are also other differences. 1) Every change in the amplitude of the physical vibrations corresponds to a subjective change in both intensity and quality, as we noted

---

[4] The same false reasoning from the names of sensations, has even led some scholars to assume that the sensation blue developed later than other color-sensations, because, for example, even in Homer the word for blue is the same as that for "dark". Tests of the color-sensations of uncivilized peoples whose languages are much more deficient in names for colors than that of the Greeks at the time of Homer, have given us a superabundance of evidence that this assumption is utterly without ground.

[5] Many physicists, to be sure, believe that an analogous relation is to be found between tones of different pitch, in the fact that every tone has in its octave a similar tone. But this similarity, as we shall see (§ 9), does not exist between simple tones, but depends on the actual sympathetic vibration of the octave in all compound clangs. Attempts to support this supposed analogy by finding in the color-line intervals corresponding to the various tonal intervals, third, fourth, fifth, etc., have all been entirely futile.

43

above in the discussion of sensations of brightness. 2) All light, even though it be made up of all the different kinds of vibration, is simple in sensation, just as much as objectively simple light, which is made up of only one kind of waves, as is immediately apparent if we make a subjective comparison of sensations of chromatic light with those of achromatic light. From the first of these facts it follows that light which is physically simple may produce not only chromatic, but also achromatic sensations, for it approaches white when the amplitude of its vibrations increases, and black when the amplitude decreases. The quality of an achromatic sensation does not, therefore, determine unequivocally its source; it may be produced either through a change in the amplitude of objective light-vibrations or through a mixture of simple vibrations of different wave-lengths. In the first case, however, there is always connected with the change in amplitude a change in the grade of brightness, which does not necessarily take place when a mixture is made.

22. Even when the grade of brightness remains constant, this achromatic sensation may have one of several sources. A sensation of pure brightness of a given intensity may result not only from a mixture of all the rates of vibration contained in solar light, as, for example, in ordinary daylight, but it may also result when only *two* kinds of light-waves, namely those which correspond to sensations subjectively the most different, that is, to opposite colors, are mixed in proper proportions. Since opposite colors, when mixed objectively, produce white, they are called *complementary* colors. As examples of such opposite or complementary colors we may mention spectral red and green-blue, orange and sky-blue, yellow and indigo-blue.

Like achromatic sensations, each of the color-sensations may also, though to a more limited extent, have one of several sources. When two objective colors which lie nearer each other in the color-circle than opposites, are mixed, the mixture appears, not white, but of a color which in the series of objectively simple qualities lies between the two with which we started. The saturation of the resulting color is, indeed, very much diminished when the components of the mixture approach opposite colors; but when they are near each other, the diminution is no longer perceptible, and the mixture and the corresponding simple color are generally subjectively alike. Thus, the orange of the spectrum is absolutely indistinguishable from a mixture of red and yellow rays. In this way, ,all the colors in the color-circle between red and green can be obtained by mixing red and green, all between green and violet by mixing green and violet, and, finally, purple, which is not in the solar spectrum, can be produced by mixing red and violet. The whole series of color-tones possible in sensation can, accordingly, be obtained from *three* objective colors. By means of the same three colors we can also produce white with its intermediate stages. The mixture of red and violet gives purple, and this is the complementary color of green; and the white secured by mixing these complementary colors, when mixed in different proportions with the various colors, gives the different grades of saturation.

23. The three objective colors that may be used in this way to produce the whole system of light-sensations, are called fundamental colors. In order to indicate their significance, a *triangular* surface is chosen to represent the system of saturation, rather than the

circular surface which is derived from the psychological relations alone. The special significance of the fundamental colors is then expressed by placing then at the angles of the triangle. Along the sides are arranged the color-tones in their maximal saturation, just as on the circumference of the color-circle, while the other grades of saturation in their transitions to white, which lies in the center, are on the triangular surface. Theoretically, any set of three colors could be chosen as fundamental colors, provided they were suitably distant from one another. Practically, those mentioned, red, green, and violet, are preferable for two reasons. First, by using them we avoid having as one of the three, purple, which can not be produced by objectively simple light. Secondly, at the two ends of the spectrum sensations vary most slowly in proportion to the period of vibration, so that when the extreme colors of the spectrum are used as fundamental colors, the result obtained by mixing two neighboring ones is most like the intermediate, objectively simple color.[6]

24. These phenomena show that in the system of light sensations a simple relation does not exist between the physical stimuli and the sensations. This can be understood from what has been said above (3) as to the, character of the *physiological* stimulation. The visual sense is to be reckoned among the *chemical* senses, and we can expect a simple relation only between the photochemical processes ill the retina and the sensations. Now, we know from experience that different kinds of physical light produce like chemical disintegrations, and this explains in general the possibility mentioned above, of having the same sensation from many different kinds of objective light. According to the principle of parallelism between changes in sensation and in the physiological stimulation, it may be assumed that the various physical stimuli which cause the same sensation all produce the same photochemical stimulation in the retina, and that altogether there are just as many kinds and varieties of the photochemical processes as kinds and varieties of distinguishable sensations. In fact, all that we know, up to the present time, about the physiological substratum of light-sensations is based upon this assumption. The investigation of the physiological processes of stimulation through light, has not yet given any further result than that the stimulation is in all probability a chemical process.

25. The relatively long *persistence of the sensation* after the stimulation that originated it, is explicable on the assumption that the light-stimulations are due to chemical processes in the retina. This persistence is called, with reference to the object used as stimulus, the *after-image* of the impression. At first this after-image appears in the same brightness and color as the object: white when the object is white, black when it is black, and if it is colored, in the same color. These are the positive and like-colored after-images. After a short time it passes, in the case of achromatic impressions, into the opposite grade of brightness, white into black, or black into white; in the case of colors, it passes into the opposite or complementary color. These are the negative and

---

[6] In the neighborhood of green this advantage does not exist, and the mixtures always appear less saturated than the intermediate simple colors. This is a clear proof that the choice of the three fundamental colors mentioned is indeed the most practical, but nevertheless arbitrary, and at bottom due to the familiar geometrical principle that a triangle is the simplest figure that can enclose a finite number of points in the same plane.

complementary after-images. If light-stimuli of short duration act upon the eye in darkness, this transition may be repeated several times. A second positive after-image follows the negative, and so on, so that an oscillation between the two phases takes place. The positive after-image may be readily explained by the fact that the photochemical disintegration caused by any kind of light, lists a short time after the action of the light. The negative and complementary after-images can be explained by the fact that disintegration in a given direction causes a partial consumption of the photochemical substance most directly concerned, and this results in a corresponding modification of the photochemical processes when the stimulation of the retina continues.

26. The origin of a part of the phenomena included under the name *light contrasts* and *color-contrasts* is very probably the same as that of the negative and complementary after-images. These phenomena consist in the appearance of simultaneous sensations of opposite brightness and color in the neighborhood of any light-impression. Thus, a white surface appears to be surrounded by a dark margin, a black surface by a bright margin, and a colored surface by a margin of the complementary color. These phenomena, which are called "marginal contrasts" when they are limited to the immediate neighborhood of the object, are in part at least nothing but negative or complementary after-images that are simultaneously visible in the immediate neighborhood of the impression as a result of continual weak ocular movements. Whether there is also an irradiation of the stimulation is a question; its existence still wants certain proof. The fact that these contrasts increase as the light becomes more intense, just as after-images do, speaks for their interconnection with the latter. In this respect, this *physiological* contrast differs essentially from certain *psychological* contrast-phenomena, with which it is generally confused. The latter are closely connected in their rise, with numerous other forms of psychological contrast, so that we will not discuss them until later, when we enter into the general treatment (§17, 9) of such phenomena.

26a. If we take the principle of parallelism between sensation and physiological stimulation as the basis of our suppositions in regard to the processes that occur in the retina, we may conclude that analogous independence in the photochemical processes corresponds to the relative independence which appears between achromatic and chromatic sensations. *Two* facts, one belonging to the subjective sensational system, the other to the objective phenomena of color-sensation can be most naturally explained on this basis. The first is the, tendency that every color-sensation shows, of passing into one of pure brightness when the grade of its brightness decreases or increases. This tendency is most simply interpreted on the assumption that every color-stimulation is made up of two physiological components, one corresponding to the chromatic, the other to the achromatic stimulation. To this assumption we may easily add the further condition, that for certain medium intensifies of the stimuli the chromatic components are the strongest, while for greater and smaller intensifies the achromatic components come more and more to the front. The second fact, is that any two opposite colors are complementary; that is, when mixed in suitable proportions, they produce an achromatic sensation. This phenomenon is most easily understood when we assume that

46

opposite colors, which are subjectively the greatest possible differences, represent objective photochemical processes that neutralize each other. The fact that as a result of this neutralization an achromatic stimulation arises, is very readily explained by the presupposition that such a stimulation accompanies every chromatic stimulation from the first, and is therefore all that is left when antagonistic chromatic stimulations counteract each other. This assumption of a relative independence between the chromatic and achromatic photochemical processes, is supported in a very striking way by the existence of an abnormity of vision, sometimes congenital, sometimes acquired through pathological changes in the retina, namely *total color-blindness.* In such cases all stimulations are, either on the whole retina or on certain parts of it, seen as pure brightness, without any admixture of color. This is an incontrovertible proof that the chromatic and achromatic stimulations are separable physiological processes.

If we apply the principle of parallelism to the *chromatic stimulation, two* facts present themselves. The first is that two colors separated by limited, short distance, when mixed give a color that is like the intermediate simple color. This indicates that color-stimulation is a process that varies with the physical stimulus, not continuously, as the tonal stimulation, but in short stages, and in such a way that the stages in red and violet are longer than in green, where the mixture of colors fairly near each other, shows the effects of complementary action. Such a non-continuous variation of the process corresponds entirely with its chemical nature, for chemical disintegration and synthesis must always have to do with *groups* of atoms or molecules. The second fact is that certain definite colors, which correspond to rather large differences in the stimuli, are subjectively opposite colors, that is, are maximal differences, and the same colors are objectively complementary, that is, mutually neutralizing, processes. Chemical processes, however, can neutralize each other only when they are in some way opposite in character. Any two complementary color-stimulations must, therefore, stand in a relation to each other similar to that which exists between the neutralizing processes operative in the case of antagonistic achromatic stimulations. Still, there are two very essential differences here. First, this opposition in the character of color-stimulations is not limited to one case, but appears for every color distinguishable in sensation, so that we must conclude, according to our presupposition, that for every stage of the photochemical process of. chromatic stimulation which is to be assumed on the ground of the results obtained by mixing neighboring colors, there is a certain complementary process. Secondly, the difference between two opposite colors, which is subjectively the greatest possible difference, is mediated by transitional forms, not merely in one direction from each color, as in the case of black and white, but in two opposite directions. In a similar way, the objective complementary action of two colors gradually diminishes as, starting from opposite colors, they approach each other in either of these two directions. We may, then, infer from this twofold elimination of complementary action that the return of the color-line to its starting point corresponds to a repetition of related photochemical processes, on the same grounds that led us to infer the opposite character of the processes corresponding to opposite colors, from the fact that they are complementary. The whole process of chromatic stimulation, beginning with red and passing beyond violet through purple mixtures to its starting point, running parallel, as it does., with continuous changes in the wavelength of objective light, is to be regarded as an indefinitely long succession of photochemical processes. All these processes together, form a closed *circle* in which, for every stage there is a neutralizing opposite and a possible transition to this opposite in two different directions.

We know nothing about the total number of photochemical stages in this circle of processes. The numerous attempts made to reduce all color-sensations to the smallest possible number of such stages, lack adequate foundation. Sometimes they indiscriminately translate the results of physical color-mixing into physiological processes, as in the assumption of *three* fundamental colors, red, green, and violet, from the different mixtures of which all sensations of light, even the achromatic, are to be derived (Young-Helmholtz hypothesis). Sometimes they start with the psychologically untenable assumption that the naming of colors is not due to the influence of certain external objects, but to the real significance of the corresponding sensations, and assume accordingly *four* fundamental colors as the sources of all color-sensations. The four fundamental colors here assumed are the two pairs red and green, yellow and blue, to which are added the similar pair of sensations of pure brightness, black and white. All other light-sensations such as grey, orange, violet, etc., are regarded as subjectively and objectively mixed colors (Hering's hypothesis). The evidence in support of the first as of the second of these hypotheses has been derived for the most part from the riot infrequent cases of *partial color-blindness.* Those who accept three fundamental colors, assert that all these cases are to be explained as a lack of the red or green sensations, or else as a lack of both. Those who accept four, hold that partial color-blindness always includes two fundamental colors that belong together as opposites, and is, therefore, either red-green-blindness or yellow-blue-blindness. An unprejudiced examination of color-blindness does not justify either of these assertions. The three-color theory can not explain total color blindness, and the four-color theory is in contradiction to cases of pure red-blindness and pure green-blindness. Finally, both theories are overthrown by the cases that unquestionably occur, in which such parts of the spectrum as do not correspond to any of the three or four fundamental colors, appear colorless. The only thing that our present knowledge justifies us in saying, is that every simple sensation of light is conditioned physiologically by a combination of two photochemical processes, a *monochromatic* and a *chromatic.* The first is made up, in turn, of a process mainly of disintegration, when the light is more intense, and a process of restitution, when the light is weaker. The chromatic process varies by stages in such a way that the whole series of photochemical color-disintegrations forms a circle *of processes* in which the products of the disintegration for any two relatively most distant stages, neutralize each other.[7]

Various changes as a result of the action of light have been observed in the living retina, all of which go to support the assumption of a photochemical process. Such are the

---

[7] The further assumption is made by the defenders of the four fundamental colors, that two opposite colors are related just as bright and dark achromatic stimulations, that is, that one of these colors is due to a photochemical disintegration (dissimilation), the other to a restitution (assimilation). This is an analogy that contradicts the actual facts. The result obtained by mixing complementary colors is on its subjective side a *suppression* of the color-sensation, while the mixture of white and black, on the other band, produces an *intermediate* sensation.

gradual change into a colorless state, of a substance which in the retina not exposed to light is red (bleaching of the visual purple); microscopical movements of the pigmented protoplasm between the sensitive elements, or rods and cones; and, finally, changes in the form of the rods and cones themselves. Attempts to use these phenomena in any way for a physiological theory of light stimulation, are certainly premature. The most probable conclusion which we can now draw is that the difference in the *forms of* the rods and cones is connected with a difference in function. The center of the retina, which is the region of direct vision in the human eye, has only cones, while in the eccentric parts the rods are more numerous; furthermore, in the center (which also wants the visual purple) the discrimination of colors is much better than in the eccentric regions, while the latter are much more sensitive to brightness. The natural conclusion from these facts is that the differences in sensitivity are connected with the photochemical properties of the rods and cones. Still, we lack here too any particular evidence.

7. SIMPLE FEELINGS

1. Simple feelings may originate in very many more ways than simple sensations, as was noted in §5. Even such feelings as we never observe except in connection with more or less complex ideational processes, have a simple character. Thus, for example, the feeling of tonal harmony, is just as simple as the feeling connected with a single tone. Several tonal sensations together are required to produce a harmony, so that it is a compound so far as its sensational contents are concerned, but the affective quality of certain harmonious compound clangs is so different from that of the feelings connected with the single tones, that both classes of feelings are, subjectively, equally irreducible. The only essential difference between the two is that the feelings which correspond to simple sensations can be easily isolated from the interconnections of which they form a part in our experience, by the same method of abstraction which we employed in discovering the simple sensations. Those, on the other hand, that are connected with some composite ideational compound, can never be separated from the feelings which enter into the compound as subjective complements of the sensations. Thus, for example, it is impossible to separate the feeling of harmony connected with the chord *c e g* from the simple feelings connected with each of the single tones *c, e*, and *g*. The latter may, indeed, be pushed into the background, for as we shall see later (§9, 3a), they always unite with the feeling of harmony to form a unitary *total feeling,* but they can never be eliminated.

2. The feeling connected with a simple sensation is commonly known as a *sense-feeling,* or the *affective tone of a sensation.* These two expressions are capable of misinterpretation in two opposite senses. There is a tendency to think that by "sense-feeling" we mean not merely a component of immediate experience that may be isolated through abstraction, but one that really exists by itself. "Affective tone", on the other

hand, may be regarded as an affective quality that must inevitably belong to a sensation, just as "color-tone" is a necessary determinant of a color-sensation. In reality, however, a sense-feeling without a sensation can no more exist than can a feeling of total harmony without tonal sensations. When, as is sometimes the case, the feelings accompanying sensations of pain, of pressure, of hot, and of cold, and muscle-sensations, are called independent sense-feelings, it is due to the confusion of the concepts sensation and feeling which is still prevalent, especially in physiology. As a result of this confusion certain sensations, such as those of touch, are called, "feelings", and in the case of some sensations accompanied by strong feelings, as sensations of pain, the discrimination of the two elements is neglected. In the second place, it would be just as inadmissible to ascribe to a given sensation a definite feeling fixed in quality and intensity. The real truth is that in every case the sensation is only one of the many factors that determine the feeling present at a given moment; besides the sensation, the processes that have gone before and the permanent dispositions - conditions that we can only partially account for in special cases - play an essential part. The concept "sense-feeling" or "affective tone" is, accordingly, in the double sense the product of analysis and abstraction: first, we must think of the simple feeling as separated from the concomitant, pure sensation, and secondly, we must pick out from among all the various changing affective elements which are connected with a given sensation under different conditions the one that is most constant and is connected with the sensation after the removal, so far as possible, of the influences, that could disturb or complicate the simple effect of the sensation.

The first of these conditions is comparatively easy to meet, if we keep in mind the psychological meaning of the concepts sensation and feeling. The second is very difficult, and, especially in the case of the most highly developed sensational systems, the auditory and visual, it is never really possible to remove entirely such *indirect* influences. We can infer what the pure affective tone of a sensation is, only by means of the same method that has already been used for the abstraction of pure sensations. Here, too, we may assume that only that affective tone which remains constant when all other conditions change, belongs to the sensation itself. The rule is easily applied to sensation, but only with great difficulty to feelings, because the secondary influences referred to are generally as closely connected with the sensation as is the primary occasion of the affective tone. Thus, for example, the sensation green arouses almost unavoidably the idea of green vegetation, and since there are connected with this idea composite feelings whose character may be entirely independent of the affective tone of the color itself, it is impossible to determine directly whether the feeling observed when a green impression is presented, is a pure affective tone, a feeling aroused by the attending idea, or a combination of both.

2a. This difficulty has led many psychologists to argue against the existence of any pure affective tone whatever. They assert that every sensation arouses some accompanying ideas and that the affective action of the sensation is due in every case to these ideas. But the results of experimental variation of the conditions for light-sensations, tell against this view. If the attendant ideas were the only sources of the feeling, it would necessarily be strongest when the sensational contents of the impression were most like those of the ideas. This is by no means the case. The affective tone of a color is greatest when its grade of saturation reaches a maximum. The pure spectral colors observed in surrounding darkness have the strongest affective tone. These colors are, however, generally very different from those of the natural objects to which accompanying feelings might refer. There is just as little justification for the attempts to derive tonal feelings from such ideas exclusively. It can not be doubted that familiar musical ideas may be aroused through a single tone; still, on the other hand, the constancy with which certain tonal qualities are chosen to express particular feelings, as, for example, deep tones to express grave and sad feelings, can be understood only under the condition that the corresponding affective quality belongs to the simple tonal sensation. The circle in which the argument moves is still more obvious when the affective tones of sensations of taste, smell, and the general sense are derived from the accompanying ideas. When, for example, the agreeable or disagreeable tone of a taste-sensation is increased by the recollection of the same impression as experienced before, this can be possible only under the condition that the earlier impression was itself agreeable or disagreeable.

3. The varieties of simple sense-feelings are exceedingly numerous. The feelings corresponding to a particular sensational system also form a system, since, in general, a change in the quality or intensity of the affective tone runs parallel to every change in the quality or intensity of the sensations.

At the same time these changes in the affective systems are essentially different from the corresponding changes in the sensational systems, so that it is impossible to regard the affective tone as a third determinant of sensations, analogous to quality and intensity. If the intensity of a sensation is varied, the affective tone may change not only in intensity, but also in quality; and if the quality of the sensation is varied, the affective tone usually changes in quality and intensity both. For example, increase the sensation sweet in intensity and it changes gradually from agreeable to disagreeable. Or, gradually substitute for a sweet sensation one of sour or bitter, keeping the intensity constant, it will be observed that, for equal intensifies, sour and, more especially, bitter produce a much stronger feeling than sweet. In general, then, *every in sensation is essentially accompanied by a twofold change in feeling.* The way in which changes in the quality and intensity of affective tones are related to each other follows the principle already stated that every series of affective changes in *one* dimension ranges between *opposites,* not, ,as is the case with the corresponding sensational changes, between greatest differences.

4. In accordance with this principle, the greatest qualitative differences in sensations correspond to the greatest opposites in affective *quality,* and to maxima of affective *intensity* which are either equal or at least approximately equal, according to the special peculiarities of the qualitative opposites. The middle point between these two opposites corresponds to an absence of all intensity, so far as only the single dimension to which the opposites belong is concerned. This absence of intensity can be observed only when the corresponding sensational system is *absolutely one-dimensional.* In all other cases, a point which is a neutral middle for one particular series of sensational

differences, belongs at the same time to another sensational dimension or even to a number of such dimensions, each of which it has a definite affective value. Thus, for example, spectral yellow and blue are opposite colors which have corresponding opposite affective tones. In passing gradually along the color-line from one of these to the other, green would be the neutral middle between them. But green itself stands in affective contrast with its opposite color, purple; and, furthermore, it is, like every saturated color, one extremity of a series made up of the transitional stages of a single color-tone to white. Again, the system of simple tonal sensations forms a continuity of only one dimension, but in this case more than in others it is impossible to isolate the corresponding affective tones through abstraction, as we did the pure sensations, because in actual experience we always have, not only intermediate stages between tones of different pitch, but also transitions between absolutely simple tones and noises made up of a profusion of simple tones. The result of these conditions is that every many-dimensional sensational system has a corresponding complex system of affective tones, in which every point generally belongs at once to several dimensions, so that the feeling corresponding to a given sensation is a resultant of the affective elements due to its position in various dimensions of the sensational system. It follows that discrimination between simple and composite feelings in the sphere of affective qualities, can not be carried out. The feeling that corresponds to a particular sensation, is as a rule, for the reasons given, a product of the fusion of several simple feelings, though it is still as irreducible as a feeling of originally simple nature (cf. §12, 3). A further consequence is that the neutral middle between opposite affective qualities, can be actually found in experience only in the special cases where the affective tone of a particular sensation corresponds to the neutral middle of all the dimensions to which it belongs. This special condition is obviously fulfilled for the many-dimensional sensational systems, especially those of sight and hearing, in just the cases in which it is of special practical value for the undisturbed occurrence of affective processes. In the one case, sensations of medium brightness and those of the low grades of chromatic saturation approximating them, in the other, the auditory impressions of our ordinary environment, which are between a tone and noise in character (as, for example, the human voice), form the neutral indifference-zones of affective quality. On both sides of these zones arise the more intense affective tones of the more marked sensational qualities. The existence of such indifference-zones makes it possible for the complex feelings which correspond to the various combinations of these, sensational qualities, to develop almost independently, without reference to the accompanying sense-feelings.

5. The variations in affective quality and intensity that run parallel to the *grades of sensational intensity,* are much simpler. They can be most clearly seen in the homogeneous sensational systems of the general sense. Each of these systems is of a uniform quality throughout, and can be fairly well represented geometrically by a single point, so that the only possible sensational changes are those of intensity, and these can be attended only by a one-dimensional series of affective changes between opposites. The neutral indifference zone is, accordingly, always easy to observe in these cases. It corresponds to the medium sensations of pressure, hot, and cold, that are connected with the normal, medium intensity of ordinary sense-stimuli. The simple feelings on both sides of this zone exhibit decidedly opposite characters, and can, in

general, be reckoned, on the one side, to pleasurable feelings, on the other, to unpleasurable (v. inf. 6). The unpleasurable feelings are the only ones that can be produced with certainty, by increasing the intensity of the sensation. Through habituation to moderate stimuli, such Iii expansion of the indifference-zone has taken place in these systems of the general sense, that when the stimuli are weak, as a rule only a succession of sensations very different in intensity or quality, can produce noticeable feelings. In such cases, feelings of pleasure always correspond to sensations of medium intensity.

The regular relation between sensational intensity and affective tone, can be better observed without this influence of contrast, in the case of certain sensations of smell and taste. At first a pleasurable feeling arises with weak sensations and increases with the increasing intensity of the sensations to a maximum, then it sinks to zero with a certain medium sensational intensity, and finally, when this intensity increases still more, the feeling becomes unpleasurable and increases until the sensational maximum is reached.

6. The variety of simple affective qualities is exceedingly great, much greater than that of, sensations. This is due to two facts. First, every sensation of the many-dimensional systems 'belongs at once to several series of feelings. Secondly, and this is the chief reason, the different compounds arising from the various combinations of sensations, such as intensive, spacial, and temporal ideas, and also certain stages in the course of emotions and volitions, have corresponding feelings, which are, as above remarked, irreducible, and must therefore be classed among the simple feelings.

It is greatly to be regretted that our names for simple feelings are so much more hazy than those for sensations. The proper nomenclature of feeling is limited entirely to the expression of certain general antitheses, as pleasurable and unpleasurable, agreeable and disagreeable, grave and gay, excited and quiet, etc. These designations are usually based on the emotions into which the feelings enter as elements, and are so general that each includes a large number of simple feelings of very different character. In other cases, complex ideas whose affective character is similar, are used in describing the feelings connected with certain simple impressions, as, for example, by Goethe in his description of the affective tone of colors, and by many musical writers in describing the feelings accompanying clangs. This poverty of language in special names for the feelings, is a psychological consequence of the subjective nature of the feelings. All the motives of practical life which give rise to the names of objects and their attributes, are here wanting. To conclude, for this reason, that there is a corresponding poverty of simple affective qualities themselves, is a gross psychological mistake, which is furthermore fatal since it makes an adequate investigation of the composite affective processes impossible from the first.

7. In consequence of the difficulties indicated, a complete list of simple affective qualities is out of the question, even more than is such a list in the case of simple sensations. Then, too, there are still other reasons why it would be impossible. The feelings, by virtue of the attributes described above, do not form closed systems, as do

the sensations of tone, of light, or of taste, but are united in a single manifold, interconnected in all its parts. Furthermore, the union of certain feelings gives rise to feelings which are not only unitary, but even simple in character. In this manifold of feelings, made up, it is, of a great variety of most delicately shaded qualities, it is nevertheless possible to distinguish certain different *chief directions*, including certain affective opposites of predominant character. Such directions may always be designated by the *two* names that indicate their opposite extremes. Each name is, however, to be looked upon as a collective name including an endless number of feelings differing from one another.

Three such chief directions may be distinguished; we will call them the direction of *pleasurable* and *unpleasurable* feelings, that of *arousing* and *subduing* (exciting and depressing) feelings, and finally that of feelings of *strain* and *relaxation*. Any concrete feeling may belong to all of these directions or only two or even only one of them. The last mentioned possibility is all that makes it possible to distinguish the different directions. The combination of different affective directions which ordinarily takes place, and the above mentioned influences which are due to the overlapping of feelings arising from various causes, all go to explain why we are perhaps never in a state entirely free from feeling, although the general nature of the feelings demands an indifference-zone.

8. Feelings connected with sensations of the general sense and with impressions of smell and taste, may be regarded as good examples of pure pleasurable and unpleasurable forms. A sensation of pain, for example, is regularly accompanied by an unpleasurable feeling without any admixture of other affective forms. In connection with pure sensations, arousing and subduing feelings may be observed best in the case of color-impressions in clang-impressions. Thus, red is arousing, blue subduing. Feelings of strain, and relaxation are always connected with the temporal course of processes. Thus, in expecting a sense-impression, we note a feeling of strain, and on the arrival of the expected event, a feeling of relaxation. Both the expectation and satisfaction may be accompanied at the same time by a feeling of excitement or, under special conditions, by pleasurable or unpleasurable feelings. Still, these other feelings may be entirely absent, and then those of strain and relaxation are recognized as specific forms which can not be reduced to others, just as the two directions mentioned before. The presence of more than one direction may be discovered in the case of very many feelings, nevertheless, simple in quality, just, as much as the feelings mentioned. Thus, the feelings of seriousness and gaiety connected with the, sensible impressions of low and high tones or dark and bright colors, are to be regarded as characteristic qualities which are outside the indifference-zone in both the pleasurable and unpleasurable direction and the exciting and depressing direction. We are never to forget here that pleasurable and unpleasurable, exciting and depressing, are not names of single affective qualities, but of *directions,* within which an indefinitely large number of simple qualities appear, so that the unpleasurable quality of seriousness is not only to be distinguished from that of a painful touch, of a dissonance, etc., but even the different cases of seriousness itself may vary in their quality. Again, the direction of pleasurable and unpleasurable feelings, is united with that of feelings of strain and relaxation, in the

case of the affective tones of rhythms. The regular succession of strain and relaxation in these cases is attended by pleasure, the disturbance of this regularity by the opposite feeling, as when we are disappointed or surprised. Then, too, under certain circumstances the feeling may, in both cages, be of an exciting or a subduing character.

9.These examples lead very naturally to the assumption that the three chief directions of simple feelings depend on the relations in which each single feeling stands to the whole *succession of psychical processes*. In this succession every feeling has in general a *threefold* significance 1) It represents a particular modification of the *state of the present moment;* this modification belongs to the *pleasurable* and *unpleasurable* direction. 2) It exercises a certain definite influence on the *succeeding* state; this influence can be distinguished in its opposite forms as *excitation* and *inhibition.* 3) It is determined in its essential character by the *preceding* state; this determining influence shows itself in the given feeling in the forms of *strain,* and *relaxation.* These conditions also render it improbable that other chief directions of feeling exist.

9a. Of the three affective directions mentioned, only that of pleasurable and unpleasurable feelings has generally been recognized; the others are reckoned as emotions. But the emotions, as we shall see in §13, come from combinations of feelings; it is obvious, therefore, that the fundamental forms of emotions must have their antecedents in the affective elements. Some psychologists have regarded pleasurable and unpleasurable feelings, not as collective terms including a great variety of simple feelings, but as entirely uniform, concrete states, so that, for example, the unpleasurableness of a toothache, of an intellectual failure, and of a tragic experience are all regarded as identical in their affective contents. Still others seek to identify the feelings with special sensations, especially with cutaneous sensations and muscle-sensations. Such entirely untenable assertions require no criticism. They indicate, however, the uncertain state of the doctrine of feelings, even at the present time.

10. The question has been raised whether or not particular *physiological processes* correspond to the simple feelings, as is the case for the sensations. Older psychology was inclined to answer this question in the negative, and to contrast the feelings as inner, purely psychological, states with sensations as processes aroused from without. In modern times, on the contrary, the affirmative answer has generally been given, but for the most part without the support of adequate empirical proof. Obviously, our assumptions in regard to the physiological phenomena accompanying the feelings must be based on actually demonstrable physiological processes, just as our assumptions in regard to the physiological conditions of sensations were deduced from the structure and functions of the sense-organs. In looking for such processes, it follows from the subjective nature of the feelings, that we should not expect to find them among the processes produced in the organism directly by external agents, as the sensations are, but rather in *reactions* which arise indirectly from these first processes. The observation of compounds made up of affective elements, that is, of emotions and volitions, whose easily perceptible concomitants are always external movements or changes in the state of the organs of movement, also points in the same direction.

The analysis of sensations, and of the psychical compounds derived from them, makes direct use of the *impression-method;* while the investigation of simple feelings, and of the processes resulting from their combinations, can employ this method only *indirectly.* On the other hand, the *expression-method,* that is, the investigation of the physiological reactions of psychical processes, is especially adapted to the examination of feelings and processes made up of them, because as shown by experience, such reactions are regular symptoms of affective processes. All the phenomena in which the inner state of the organism is outwardly expressed, may be utilized as aids for the expression-method. Such are, besides the movements of the external muscles, especially the respiratory and cardiac movements, the contraction and dilation of the blood-vessels in particular organs, the dilation and contraction of the pupil of the eye, etc. The most delicate of these is the beating of the heart, which can be examined as exactly reproduced in the pulse of some peripheral artery. All other phenomena are generally wanting in the case of a simple feeling. It is only for high intensifies, where the feelings always pass into emotions, that we have other, added symptoms, especially changes in respiration, and mimetic expressive movements.

11. Of the chief directions of feeling mentioned above, especially that of *pleasarable and unpleasurable* feelings can be shown to stand in regular relation to the pulse. When the feeling is pleasurable, the pulse is retarded and intensified, when unpleasurable, the pulse is accelerated and weakened. For the other directions, the accompanying changes can only be inferred with some degree of probability, from the effects of the corresponding emotions (§13, 5). Thus, *exciting* feelings seem to betray their presence only through stronger pulse beats, and *subduing* through weaker, without a change of rate in either case. For feelings of *strain,* we have retarded and weakened pulse, for those of *relaxation,* accelerated and intensified pulse. Single feelings belong for the most part to several of these directions at the same time; as a result, the action of the pulse is in many cases so complex that the most that can be concluded is the predominance of one or the other direction. The conclusion is, however, uncertain so long as it is not confirmed by direct observation of the feeling.

11a. The relations that seem probable from experiments on the symptoms of feelings and emotions as found in pulse-activity, may be presented in the following scheme.

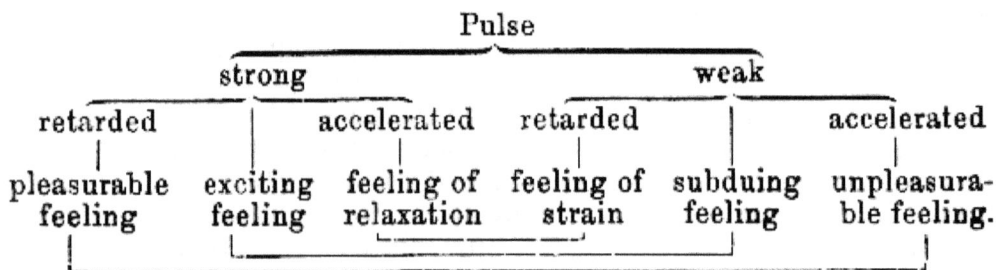

Exciting and subduing feelings, then, show themselves by simple changes in the pulse, the others by double changes. But this scheme, which is derived for the most part from the effect of complex emotions, needs confirmation from experiments in which attention is paid to the isolation of these various affective directions. Changes in respiration, muscle-tension, etc., also need further investigation. It is obvious from the equivocal character of each symptom, that when a particular feeling is given in psychical experience, we can infer particular resulting innervations from the symptoms which appear, but that we can never infer the presence of particular feelings from the physiological symptoms. It follows that the expression-method can not be as highly valued from a psychological point of view as the impression-method. >From the very nature of the case, the impression-method is the only one that can be used in arousing and varying psychical processes at will. The expression-method gives results that explain only the physiological phenomena which accompany feelings, not the psychological nature of the feelings themselves.

The variations observed in the pulse must be regarded as the results of a changed innervation of the heart, coming from the cardiac center in the brain. Physiology shows that the heart is connected with the central organs by two kinds of nerves: *excitatory nerves,* which run through the sympathetic system and originate indirectly in the medulla, and *inhibitory nerves,* which belong to the tenth cranial nerve (vagus) and also have their source in the medulla. The normal regularity of the pulse depends on a certain equilibrium between excitatory and inhibitory influences. Such influences come not only from the brain, but from the centers in the ganglia of the heart itself. Thus, every increase and every decrease of the heart's energy may be interpreted in two different ways. The first may be due to an increase of excitatory, or to a decrease of inhibitory innervation, and the second may be due to a decrease in excitatory or to an increase in inhibitory innervation, or in both cases the two influences may be united. We have no universally applicable means of investigating these possibilities, still, the circumstance that the stimulation of the inhibitory nerves has a quicker effect than the stimulation of the excitatory, gives us good ground in many cases for conjecturing the presence of the one or the other. Now, the changes in the pulse always follow very quickly the sensations that cause them. It is, therefore, probable that in the case of feelings and emotions, we have chiefly changes in inhibitory innervation, originating in the brain and conducted along the vagus. It may well be assumed that the affective tone of a sensation on its physiological side, corresponds to a spreading of the stimulation from the sensory center to other central regions which are connected with the sources of the inhibitory nerves of the heart. Which central regions are thus affected, we do not know. But the circumstance that the physiological substrata for all the elements of our psychological experience, are in all probability to be found in the cerebral cortex, leads very naturally to the assumption that the same is true for the center of these inhibitory innervations. Furthermore, the essential differences between the attributes of feelings and those of sensations, make it probable that this center is not identical with the sensory centers. If a special cortical region is assumed as the medium for these effects, there is no reason for supposing a special one for each sensory center, but the complete uniformity in the physiological symptoms goes more to show that there is only one such region, which must then at the same time serve as a kind of central organ for the connection of the various sensory centers. (For the further significance of such a central region, and its probable anatomical position, compare §15, 2a.).

CHAPTER 2

PSYCHICAL COMPOUNDS

8. Definition and Classification of Psychical Compounds

1. By "psychical compound", we mean any composite component of our composite experience which is marked off from the, other contents of this experience by particular characteristics, in such a way that it is apprehended as a relatively independent unity, and is, when necessity demands it, designated by a special name. In developing these names, language has followed the general rule that *only classes* and the most important *species* into which phenomena may be grouped, shall have special designations, while the, discrimination of concrete compounds is left to immediate perception. Thus, such expressions as ideas, emotions, volitional acts, etc., designate general classes of psychical compounds, such expressions as visual ideas, joy, anger, hope, etc., special species included in these classes. So far as these designations, which have, arisen from practical experience, are based upon actual distinguishing characteristics, they may be retained by science. But science must give an account of the nature of these characteristics and also of the peculiar contents of each of the chief forms of psychical compounds, in order to give every single an exact meaning. In doing this, we must avoid from the first *two* presuppositions to which the existence of these names might easily mislead us. The first is the view that a psychical compound is an absolutely independent content of immediate experience. The second is the opinion that certain compounds, for example, ideas, have the nature of *things.* The truth is that these compounds are only *relatively* independent units. Just as they are made up of various elements, so they themselves unite to form a complete interconnection, in which relatively simple compounds may continually combine to form more composite ones. Then, again, compounds, like the cyclical elements contained in them, are never things, but *processes* which change from moment to moment, so that it is only through deliberate abstraction, which is, indeed, indispensable for the investigation in many cases, that they can be thought of as constant at any moment.

2. All psychical compounds may be resolved into psychical elements, that is, into pure sensations and simple feelings. The two kinds of elements behave, however, in an essentially different manner, in accordance with the peculiar properties of simple feelings as described in §7. The sensational elements found by such a resolution, always belong to one of the sensational systems already considered. The affective elements, on the other hand, include not only those which correspond to the pure sensations contained in the compounds, but also those due to the interconnection of the elements

into the systems of sensational qualities, accordingly, always remain the same, no matter how great a variety of compounds arises, while the systems of simple affective qualities continually increase. Connected with this increase is another attribute which is thoroughly characteristic for the actual nature of psychical processes. The attributes of psychical compounds are never limited to those of the elements that enter into them, but new attributes, peculiar to the compounds themselves, always arise as a result of the combination of these elements. Thus, a visual idea has not only the attributes of the light-sensations and of the sensations of ocular position and movements contained in it, but also that of the spacial arrangement of the sensations, which these elements in themselves do not have. Again, a volition is not only made up of the ideas and feelings into which its single acts may be resolved, but there result from the combination of these acts, new affective elements which are specifically characteristic of the complex volition. Here, again, the combinations of sensational and affective elements are different. In the first case, on account of the constancy of the sensational systems, no new sensations can arise, but only peculiar *forms of their arrangement.* These forms are the *extensive spacial* and *temporal manifolds.* When, on the other hand, affective elements combine, *new simple feelings* arise, which unite with those originally present to make *intensive* affective units of composite character.

3. The classification of psychical compounds is naturally based upon the character of the elements that make them up. Those composed entirely or chiefly of sensations are called *ideas,* those consisting mainly of affective elements, *affective processes.* The same limitations hold here as in the case of the corresponding elements. Although compounds are more the products of immediate discrimination among actual psychical processes than the elements are, still, there is at bottom no pure ideational process and no pure affective process, but in both cases we can only abstract to a certain extent from one or the other component. As in the case of the two kinds of elements, so here we can neglect the accompanying subjective states when dealing with ideas, but must always presuppose some idea for the affective processes. Still, these ideas may be of very different kinds for the single species and varieties of affective processes.

We distinguish, accordingly, three chief forms of *ideas:* 1) intensive ideas, 2) spacial ideas, 3) temporal ideas; and three forms of *affective processes:* 1) intensive affective combinations, 2) emotions, 3) volitions. Temporal ideas constitute a sort of link between the two kinds of processes, for certain feeling play an important part in their formation.

9. INTENSIVE IDEAS

1. A combination of sensations in which every element is connected with any second element in exactly the same way as with any other, is called an intensive idea. Thus, for example, a compound clang made up of the tones *d f a* is such an intensive idea. For the immediate apprehension, each of the partial combinations into which this compound

clang can be resolved, as *df, da, fd, fa, ad, af,* are all entirely equivalent, in whatever order they are thought of. This is obvious at once if we compare the compound clang with any succession of the same tones, where *df, da, fd, fa,* etc., are essentially different ideas. We may define intensive ideas, accordingly, as *combinations of sensational elements, in which the order of the elements may be infinitely varied.*

It follows from their nature, that intensive ideas do not have, arising from the way in which their elements axe united any characteristics, by means of which they can be resolved into separate parts. Such a resolution is possible only through the differences in the constituent elements themselves. Thus, we discriminate the elements of the compound clang *d f a,* only because we hear in it the qualitatively different tones d, f, and a. Still, the separate components in such a unitary idea are less clearly distinguishable than in their isolated state. This fact, that the elements are pushed into the 'background by the impression of the whole, is of great importance for all forms of ideational combination. We call it the *fusion of sensations,* and in particular, for intensive ideas, *intensive fusion.* If the connection of one element with others is so close that it can be perceived as a part of the whole only through unusual concentration of the attention aided by experimental variation of the conditions, we call the fusion *complete.* If, on the other hand, the elements are immediately recognized in their proper qualities, and merely recede somewhat into the background in comparison with the impression of the whole, we call the fusion *incomplete.* If certain particular elements are more prominent in their characteristic qualities thin others, we call them the *predominating elements.* The concept of fusion as here defined as a *psychological* concept. It presupposes that the fused elements of the idea are really subjectively distinguishable. It must not be confounded with the entirely different and purely physiological concept of the fusion of external impressions into a single resultant stimulation. For example, when complementary colors unite and give white, the fusion is, of course, not psychological.

In reality, every intensive idea always enters into certain spacial and temporal combinations. Thus, for example, a compound clang is always a process having a certain duration, and is at the same time localized by us in some direction or other, though often only very indefinitely. But since these temporal and spacial attributes can be indefinitely varied, while the intensive character of the ideas remain the same, we may abstract from the former in investigating the intensive attributes.

2. Among *ideas of the general sense* we have intensive fusions in the form of combinations of sensations of pressure with those of hot or cold, or combinations of pain-sensations with those of temperature or pressure. All these fusions are incomplete, and very often there is no decidedly predominating element. The combination of certain sensations of *smell* and *taste* are more intimate. This is obviously favored on the physiological side by the proximity of the sense-organs, on the physical side by the regular connection between certain stimulations of the two senses. In such cases the more intense sensations are generally the predominating elements, and when these are the sensation of taste, the composite impression is usually regarded as a taste-quality

only. Thus, most of the impressions known in ordinary life as "tastes", are in reality combinations of tastes and smells.

The greatest variety of intensive ideas, in all possible gradations of complexity, are presented by the *sense of hearing*. The relatively most simple of these ideas and those which are most closely related to simple tones, are the *single clangs*. As more complex forms, we have *compound clangs. Complex noises* may arise from the latter when they are united with sensations of simple noises, and under certain other circumstances.

3. A *single clang* is an intensive idea which is made up of a series of tonal sensations regularly graded in quality. These elements, the *partial tones* of the clang, form a complete fusion, in which the sensation of the lowest partial tone becomes the predominating element. The *pitch* of the tone is determined by this *principal* tone. The other elements are higher and are, accordingly, called *overtones*. The overtones are all grouped together under the name clang-color as a second determinant of the clang, added to the predominating tone. All the partial tones that go to determine the clang color are placed along the tonal line at certain regular intervals from the principal tone. The complete series of possible overtones in a clang consists of the first octave of the principal tone, the fifth of this octave, the second octave of the principal tone, and the major third and the fifth of this second octave, etc. This series corresponds to the following proportions between the number of objective tonal waves: 1 (principal tone), 2, 3, 4, 5, 6, 7, 8, ... (overtones). When the pitch of the principal tone remains constant, only the second determinant of the tonal quality, the *clang-color,* can vary according to the number, position, and relative intensity of the overtones. In this way we can explain the great variety of clang-colors in musical instruments, as well as the fact that for every instrument the clang-color changes somewhat with the pitch; for in the case of low tones the overtones are generally relatively strong, in that of high relatively weak, while they disappear entirely when they are too high to be audible. Even the slight differences in clang-color in single instruments of the same kind, are to be explained in the same way.

From a psychological point of view the chief condition for the rise of a single clang, is the complete, or approximately complete, fusion of several tonal sensations with only *one* predominating element. As a rule, it is impossible to distinguish with the unaied ear the overtones in a clang. They can be made perceptible by the use of resonators (resonator-tubes tuned to the overtones sought), and after they have been isolated in this experimental way, the stronger ones can be successively heard in the clang, even without the aid of the resonators, if the attention is directed to them.

4. There are three conditions necessary if there is to be only *one* predominating element in a tonal fusion. First, one tone must be relatively more intense. Secondly, in its qualitative relations to the other partial tones, the principal tone must be the fundamental of a series whose members are all harmonious. Thirdly, all the partial tones must be uniformly coincident. This coincidence is objectively guaranteed by deriving the clang from a unitary source, (that is, producing the clang through the vibrations of *one* string, *one* reed-pipe, etc.) The result is that the objective vibrations of

the partial tones always stand in the same relation to one another - a result which can not be secured when clangs from several sources are united. The first two of these conditions relate to the *elements,* the third to the *form* of their combinations. The first is the least essential to the idea of a single clang. If the second is not fulfilled, the combination becomes a *compound clang* when the predominating fundamental is wanting, or a noise when the series of tones is not harmonious, or a mixed form between a clang and a noise when both parts of the condition are unfulfilled. If the third condition, of constancy in the phases of the partial tones, is not met, the clang becomes compound even when the first two conditions are complied with. A series of simple clangs from a number of tuning-forks which should unite to a single clang so far as intensity and quality are concerned, always produces in reality the idea of a compound clang.[8]

5. A *compound clang* is an intensive combination of single clangs. It is in general an incomplete fusion with several predominating elements. There are, as a rule, all possible grades of fusion in a compound clang, especially when it is made up of single clangs of composite quality. In such a case, not only does every single clang form a complete fusion in itself, but these single clangs fuse the more completely with one another the more their fundamentals approach the relation of elements of a single clang. So it comes that in a compound clang made up of single clangs rich in overtones, those components whose fundamentals correspond to the overtones of some other single clang in the compound, fuse more completely with this related clang than with others. The other clangs, in turn, fuse the more completely the more their relation approaches that of the first members of a series of overtones. Thus, in the compound clang c e g c' the clangs c and c' form a nearly complete fusion, while the fusions of the clangs c and g, c and e, are incomplete. Still less complete is the fusion between c and e^b. A measure for the degree of fusion may be obtained in all these cases by allowing an observer to hear the compound clang for a very brief interval, after which he is to decide whether he perceived only one clang or several. This experiment is repeated many times, and the relative number of judgments in favor of the unity of the clang is a measure for the degree of fusion.

6. Besides the elements contained in the single clangs of a compound, there are always, arising from the combination of vibrations in the auditory organ, additional elements which cause new tonal sensations, characteristic for the different kinds of compound clangs. These may also fuse more or less completely with the original clang. They are sensations of *difference-tones;* they correspond, as their name indicates, to the

---

[8] The case is different when the fundamental itself contains overtones of noticeable intensity, which are also repeated as independent clangs in the compound tone. The single clangs of such a series arrange themselves in the same phases as these overtones, and the compound clang has the character of a single clang with very strong overtones. Helmholtz concluded from experiments in which he combined in various ways simple clangs from tuning-forks, that differences in phase have no influence on the clang-color. But as the idea of a single clang can not be produced in this way, it is probable that an entirely constant relation of the phases of different tonal vibrations from independent sources can never be brought about with this method. Experiments by R. Koenig tell for the influence on the clang-color, of the form of the clang as determined by the relation of the vibration-phases.

difference between the number of vibrations in two primary tones. They may have a twofold origin, either from the interference of the vibrations in the outer ear, especially in the tympanum or chain of ossicles (Helmholtz's combination-tones), or from the interference of the vibrations in the auditory nerve-fibers (Koenig's beat-tones). The first are, from the very character of their origin, weak tones; especially in comparison with the original tones, they are always relatively very weak. The second class, on the other hand, are generally stronger and may even surpass the original tones in intensity. It is probable that the first appear only in the case of harmonious compound clangs, while the second appear also in dissonant compound clangs. The fusion of difference-tones with the chief tones of the compound is the more complete the less intense the former are, and the more they tend to form a simple harmonious tonal series with the original components of the clang. As a result of these attributes, the difference-tones are to compound clangs what the overtones are to single clangs. They are, however, almost entirely independent of the clang-color of the components of the compound, but vary greatly with the relation in which the principal tones of these components stand to one another. This explains the relative uniformity in the character of a given compound clang even when the clang-colors of its components vary.

7. A compound clang may pass through all possible intermediate stages into a third form of intensive auditory ideas, that of *noises*. When two tones are no longer included within a series of harmonious tones and when at the same time the difference between the number of their vibrations does not exceed certain limits, for higher tones about sixty vibrations and for lower thirty or even fewer, there arise disturbances in the compound clang, which correspond in number to the difference between the number of vibrations in the primary tones, and are due to the alternating coincidence of like and opposite phases of vibration. These disturbances are either interruptions of the clang-sensation, *beats*, or, especially in the case of deep tones, intermittent sensations of a difference-tone, *tonal beats*. If the differences in the number of vibrations exceed the numbers mentioned, the tones at first sound continuous, for the interruptions disappear, but they are harsh. Later the harshness disappears and we have *pure dissonance*. Ordinary dissonance is made up of a mixture of beats or harshness and pure dissonance. The first two are due to perceptible or just disappearing interruptions of the sensation, the latter to the entire absence of the unity of the clang, that is, of the consonance that would have arisen if a complete or partial fusion had taken place. This lack of accord in tones, due to the relation of their pure qualities, may be designated *bissonance*. If through the simultaneous sounding of a great number of non-accordant tones the various conditions for an ordinary dissonance, beats, tonal beats, harshness, and bissonance, are all added together, a *noise* is the result. On the psychological side this means that the predominating tonal elements disappear entirely or become mere modifying elements in the total idea. For our apprehension of noises, in the case of those which last a short interval only, the general pitch of the most intense elements is determinative, in the case of those which last longer, the form of the disturbance resulting from the rapidity of the beats, from the accompanying tonal beats, etc., also has an influence.

Human articulations are characteristic examples of different forms of noise. The vowels are intermediate between clangs and noises with predominantly clang character; the resonants are noises of long duration, and the proper consonants noises of short duration. In whispers the vowels become simply noises. The circumstance that the differences in vowels are perfectly distinct in whispers, goes to prove that the character of vowels depends essentially on their noise-elements. It is probable that simple sensations of noise enter into all noises together with the numerous tonal elements that no to make them up. The irregular air-vibrations arising from the disturbances in the tonal waves, excite both the nervous elements in the vestibule of the labyrinth, which are sensitive to such stimulations, and the auditory nerve fibers themselves.

7a. Helmholtz's *resonance hypothesis* has aided us materially in understanding the physiological substratum of intensive auditory ideas, especially those of clangs. It is assumed that certain parts of the auditory organ are so tuned that tonal waves of a given rate always set in sympathetic vibration only the part, correspondingly tuned. This explains in a general way the analyzing ability of the auditory sense, as a result of which we can distinguish the elements not only in a compound clang, but to some extent even in a single clang. The resonance hypothesis, however, accounts physiologically for only one side of tonal fusion. the persistence of the single sensation in the total intensive idea, not for the other side, the more or less intimate combination of the elements. The assumption of an imaginary "organ of fusion" in the brain for this purpose, is one of those fictions that are more harmful than helpful, in which the attempt is made to satisfy a demand for explanation with an empty word. The tonal elements that produce an intensive clang-idea persist as real sensations and still give up their independence more or less in the total idea. Tonal fusion is, then, a psychical process and requires a psychological explanation. But since this fusion is very different under different objective conditions, as, for example, when the impressions are due to the combined vibrations from a single source or to vibrations from several distinct sources; these differences must have some physiological and physical grounds for their explanation. The most natural way to attempt such an explanation is properly to supplement the resonance hypothesis. If we assume that besides the analyzing parts of the auditory organ, the resonant membrane, still others exist which are effected by the total, unresolved clang, we have a sufficient physiological substratum for the different effects of the various conditions. The observations on birds deprived of their labyrinths make it possible to infer that the auditory nerve-fibers in the canals of the labyrinth may be such organs. Then, too, the existence of beat-tones, which sometimes surpass the primary tones in intensity, and the observation that the interruptions of a single tone may unite to form a second sensation when sufficiently rapid, both seem to require a similar supplementation of the resonance hypothesis.

10. SPACIAL IDEAS

1. Spacial and temporal ideas are immediately distinguished from intensive ideas by the fact that their parts are united, not in an arbitrarily variable, but in a definitely fixed order, so that when the order is thought of as changed the idea itself changes. Ideas with such a fixed arrangement are called in general *extensive* ideas.

Of the possible forms of extensive ideas, *spacial* ideas are distinguished by the fact that the fixed arrangement of the parts of a spacial compound holds only for the *relation of the parts to one* another, not for their relation to the ideating subject. This latter relation may be thought of as indefinitely varied. The objective independence of spacial compounds from the ideating subject is called the *movability* and *torsibility of spacial compounds.* The number of directions in which such movement and torsion may take place; is limited. They may all be reduced to three dimensions, in each of which it is possible to advance in two opposite directions. The number of directions in which the parts of a single compound may be arranged as well as the number in which various compounds may be arranged with reference to one another, is the same as the maximal number of directions in which movement and torsion are possible. This is what we call the *three-dimensional* character of space. A single spacial idea may, accordingly, be defined as a *three dimensional compound whose parts are fixed in their location with regard to one another, but capable of indefinite variation in their location with regard to the ideating subject.* This definition neglects, of course, the frequent changes in the arrangement of the parts, which occur in reality. When these changes take place, they are to be regarded as transitions from one idea to another. This three-dimensional arrangement of spacial ideas must of necessity include one-dimensional and two-dimensional arrangements as special cases. In such cases, however, the wanting dimensions must always be added in thought as soon as the relation of the idea to the ideating subject is taken into account.

2. This relation to the ideating subject, which is really present in all spacial ideas, renders it from the first psychologically impossible that the arrangement of the elements in such an idea should be an original attribute of the elements themselves, analogous to the intensity or quality of sensations; it requires rather that this arrangement should result from the bringing together of these elements, and should arise from some new psychical conditions that come with this coexistence. If this is not admitted, it becomes necessary not only to attribute a spacial quality to every single sensation, but also to postulate for every sensation, however limited, a simultaneous idea of the whole of three-dimensional space in its location with regard to the ideating subject. This would lead to the acceptance of an a priori space-perception prior to all concrete sensations, which is not only contradictory to all our experiences as to the conditions for the rise and development of psychical compounds in general, but also contradictory to all our experiences as to the influences that affect spacial ideas in particular.

3. All spacial ideas are arrangements either of *tactical* or of *visual* sensations. Indirectly, through of connection of other sensations with either tactual or visual ideas, the spacial relation may be carried over to other sensations. In the cases of touch and sight, it is obvious that the extended surface of the peripheral sense-organs, and their equipment with organs of movement, which render possible a varying location of the impressions in regard to the ideating subject, are both favorable conditions for an extensive, spacial arrangement of the sensations. The *tactual* sense is the earlier of the two here in question, for it appears earlier in the development of organisms and shows the structural relations in much coarser, but for that reason in many respects much plainer, form than the more delicately organized visual organ does. Still, it is to be noted that where vision is present, the spacial ideas from touch are greatly influenced by those from sight.

A. SPACIAL TOUCH-IDEAS

4. The *simplest* possible touch-idea is a *single, approximately punctiform impression on the skin.* If such an impression is presented even when the eyes are turned away, there arises a definite idea of the *place touched.* Introspection shows that this idea, which is called the *localization of the stimulus,* under the normal condition where vision is present is not immediate, as we should expect it to be if the spacial quality were an original attribute of sensations, but that it depends upon a secondary, generally very obscure, *visual idea* of the region touched. Localization is, therefore, more exact near bounding lines of the touch-organs than on the uniform intervening surfaces, since these bounding lines are more prominent in the visual images. The arousal of a visual idea through the tactual impression, even when the eyes are turned away, is possible because every point of the organ of touch gives to the touch-sensation a peculiar qualitative coloring, which is independent of the quality of the external impression and is probably due to the character of the structure of the skin, which varies from point to point and is never exactly the same in two separate regions.

This local coloring is called the *local sign* of the sensation. It varies in different regions of the skin at very different rates: rapidly on the tip of the tongue, on the ends of the fingers, and on the lips; slowly on the broader surfaces of the limbs and trunk. A measure for the rate of this variation may be obtained by applying two impressions near each other to any region of the skin. So long as the distance of the impressions is less than that of distinguishable local signs they are perceived as a single one, but so soon as they pass this limit they are perceived as specially separate. The smallest, just noticeable distance between two impressions is called the *space-threshold for touch.* It varies from one or two millimeters (tips of tongue and fingers) to sixty-eight millimeters (back, upper arm, and leg). On the pressure-spots, when the stimuli are favorably applied, still shorter distances can be perceived. Then, too, the threshold is dependent on the condition of the tactual organ and on practice. As a result of the first, for example, the threshold is smaller for children than for adults, since the differences in structure that condition the local signs, are obviously more crowded together. As a result of practice, the threshold is smaller in the case of the blind than of those who have vision, especially at the ends of the fingers, which are most used for touching.

5. The influence of visual ideas of the regions touched, where vision is present, as just described, teaches that the localization of tactual impressions and the spacial arrangement of a number of such impressions is not due to an original spacial quality of cutaneous points or to any primary space-forming function of the tactual organ. On the contrary, it presupposes spacial ideas of sight, which can be made use of, however, only because the various parts of the tactual organ have certain qualitative attributes, local signs, which arouse the visual image of the part touched. There is no reason for attributing an immediate spacial relation to the local signs themselves; it is obviously enough that they act as qualitative signals to arouse the appropriate visual images. This connection with vision depends upon the frequent union of the two. The steeliness of localization will, therefore, be aided by all the influences that increase either the clearness of the visual images or the qualitative differences in local signs.

We may describe the formation of spacial ideas in this case as the arrangement of tactical stimuli in visual images are already present. The whole process is a consequence of the constant connection of these images with the qualitative local signs of the stimuli. The union of the local signs and the visual images of the corresponding region may, then, be rewarded as an *incomplete, but very constant, fusion.* The fusion is incomplete because both visual image and tactual impression retain their independent character; but it is so constant that, when the state of the tactual organ remains the same, it seems invariable. This last fact explains the relative certainty of localization. The predominating elements of this fusion are the tactual sensations. For many persons the visual images are pushed so far into the background that they can not be perceived with any certainty, even with the greatest attention. The apprehension of space, in such cases, is perhaps an immediate function of tactual and motor sensations, as for the blind (v. inf. 6). As a rule, however, more careful observation shows that it is possible to recognize the position, and distance of the impressions only by attempting to make the indefinite visual image of the region touched more distinct.

6. The conditions that hold when vision is present, are essentially different from those found in cases of *blindness,* especially congenital or early acquired blindness. Blind persons retain for a long time memory images of familiar visual objects, so that the spacial ideas of touch always remain, to some extent, products of a fusion between tactual sensations and visual images. But these visual images can not be continually renewed, so that the persons in question come more and more to make use of movements. The *sensations of movement* that arise from the joints and muscles in passing from one tactual impression to another, serve as a measure for the movement executed and, at the same time, as a measure for the distance between the two impressions. These sensations of movement, which in acquired blindness are additions to the gradually fading visual images and in part substitutes for them, are, in congenital blindness, the only means present from a the first for the formation of an idea of the relative position and distance of the single impressions. We observe in the latter case continual movements of the touch-organs, especially the fingers, over the object. Added

to these movements are a more concentrated attention to tactual sensations and a greater practice in their discrimination. Still, the low grade of development of this sense, in comparison with sight, always shows itself in the fact that the apprehension of continuous lines and surfaces is much less perfect than that of approximately punctiform impressions arranged in various ways. The necessity of making a *blind-alphabet* of arbitrary figures formed by various combinations of raised points, is a striking proof of this. Thus, for example, in the ordinary alphabet (Braille's) one point represents A, two points in a horizontal line B, two points in a vertical line C, etc. With six points at most all the letters can be formed, but the points must be far enough apart to be perceived is separate with the end of the index finger. The way in which this alphabet is read is characteristic for the development of the space-ideas of the blind. As a rule the index fingers of both hands are used for this purpose. The right finger precedes and apprehends a group of points simultaneously (synthetic touch), the left finger follows somewhat more slowly and apprehends the single points successively (analytic touch). Both the synthetic and analytic impressions are united and referred to the same object. This method of procedure shows clearly that the spacial discrimination of tactual impressions is no more immediately given in this case than in the case where vision was present, but that here the improvements by means of which the finger that is used for analytic touch passes from point to point, play the same part as the accompanying visual ideas did in the normal cases with vision.

An idea of the extent and direction of these movements can arise only under the condition that every movement is accompanied by an inner tactual sensation. The assumption that these inner tactual sensations are immediately connected with an idea of the space passed through in the movement, would be highly improbable, for it would not only presuppose the existence of a connate perception of surrounding space and of the position of the subject in respect to the same, but it would include another particular assumption. This is the assumption that inner and outer touch-sensations, although they are otherwise alike in quality and physiological substrata, still differ in that inner sensations give, along with the sensation, an image of the position of the subject and of the spacial arrangement of the immediate environment. This would really necessitate a return to the Platonic doctrine of the memory of innate ideas, for the sensations of movements arising from touch are here thought of as the mere external occasional causes for the revival of innate transcendental ideas of space.

7. Apart from its psychological improbability, such an hypothesis as that just mentioned can not be reconciled with the influence exercised by practice on the discrimination of local signs and of differences in movements. There is no other way except to attribute the rise of spacial ideas here, as in normal cases with vision, to the *combinations of the sensations themselves as presented in experience.* These combinations consist in the fact that in passing from one outer tactual impression to another, any two sensations, *a* and *b*, with a certain difference in local signs, always have a corresponding inner touch-sensation,   , accompanying the movement; while two sensations, *a* and *c*, with a greater difference in local signs, have a more intense sensation of movement,   . For the blind there is always such a regular combination of inner and outer touch-sensations. From the strictly empirical point of view it can not be affirmed that either of these

sensational systems, itself, brings the idea of spacial arrangement; we can only say that this arrangement results regularly from the combination of the two. On this basis the special ideas of the blind, arising, as they do, from external impressions, are defined as the product of the *fusion of external tactual sensations and their qualitatively guided local signs, with internal tactual sensations, graded according to intensity.* The external sensations with their attributes as determined by the external stimulus, are the predominating elements in this fusion. These push the local signs with their qualitative peculiarities, and the sensations of movement with their intensive attributes, so far into the background, that, like the overtones of a clang they can be perceived only when the attention is especially concentrated upon them. Spacial ideas from touch are, accordingly, due to a *complete* fusion. Their characteristic peculiarity, in contrast, for example, with intensive tonal fusions, is that the subordinate and supplementary elements are different in character, and at the same time related to one another according to definite laws. They are different, for the local signs form a pure qualitative system, while the inner touch-sensations which accompany the movements of the tactual organs, form a series of intensifies. They are related, for the motor energy used in passing through an interval between two points, increases with the extent of the interval, that, in proportion to the qualitative difference between the local signs, there must also be an increase in the intensity of the sensations of movement.

8. The spacial arrangement of tactual impressions is thus the product of a *twofold fusion*. First, the subordinate elements fuse, in that the various qualities of the local sign system, which is spread out in two dimensions, are related to one another according to the grades of intensity of the sensations of movement. Secondly, the tactual impressions as determined by the external stimuli, fuse with the product of the first union. Of course, the two processes do not take place successively, but in one and the same act, for the local signs and movements must both be aroused by the external stimuli. Still, the external sensations vary with the nature of the objective stimulus, while the local signs and internal tactual sensations are subjective elements, whose mutual relations always remain the same even when the external impressions vary. This is the psychological condition for the *constancy of attributes* which we ascribe to space itself, in contrast wich [sic] the great changeableness of the qualitative attributes of objects in space.

9. After the fusion between local signs and internal tactual sensations has once been effected, either one of these elements lay itself, though perhaps in a limited degree, is able to bring out a localization of the sensations, and even to arouse complex spacial ideas. In this way not only normal individuals with vision, but also the blind, even the congenitally blind, have an idea of the place touched, and can perceive as specially separate two impressions that are far enough apart, even when the touch-organs remain perfectly quiet. Of course, the congenitally blind can have no visual image of the region touched, but they have instead of this an idea of a movement of the part touched and, where several impressions are received, the idea of a movement from one to the other. The same fusion takes place in ideas thus formed as in the ordinary ones, where movements are really present, only here the one factor, the inner tactual sensation is merely a memory-image.

10. In the same way, we may have the reverse process. The real contents of experience may be a sum of inner tactual sensations which arise from the movement of some part of the body, while no noticeable external tactual sensations whatever are given, and yet these external sensations which accompany the movement may still be the basis of a spacial idea. This is regularly the case when we have *pure ideas of our own movements.* If, for example, we shut our eyes and then raise our arm, we have at every moment an idea of the position of the arm. To be sure, external tactual sensations that arise from the torsion and folding of the skin, play some part here too, but they are relatively unimportant in comparison with the internal sensations from the joints, tendons, and muscles.

It can be easily observed that where vision is present, this idea of position comes from an obscure visual image, of the limb with its surroundings, which is aroused even when the eyes are closed or turned away. This connection is so close that it may arise between the mere memory-image of the inner tactual sensation and the corresponding visual idea, as is observed in the case of paralytics, where sometimes the mere will to execute a certain movement arouses the idea of a movement really executed. Evidently the ideas of one's own movements depend, when vision is present, on incomplete fusions, just as the external spacial ideas of touch do, only here the internal sensations play the part that the outer sensations play in the former case. This leads to the assumption that the inner tactual sensations also have local signs, that is, the sensations in the various joints, tendons, and muscles show certain series of local differences. Introspection seems to confirm this view. If we move alternately the knee-joint, hip-joint, and shoulder-joint, or even the corresponding joints on the right and left sides, the quality of the sensation seems each time a little different, even if we neglect the connection with a visual image of the limb, which can never be entirely suppressed. Then, too, it is impossible to see how accompanying visual images could arise at all without such differences. That would require not only a connate idea of space in the mind, but also a connate knowledge of the position and movements of the limbs in space for every moment.

11. From the relations that exist in the normal cases with vision, we can understand the way in which the ideas of their own movements arise in the case of the congenitally blind. Here, instead of a fusion with a visual image, there must be a fusion of sensations of movement with the local signs. Outer tactual sensations also act is aids in this case. In fact, they are much more important here than when vision is present. The ideas of the blind as to their own movements are exceedingly uncertain so long as they are unaided by contact with external objects. When, however, they touch such objects, they have the advantage of greater practice with the external tact sense and a keener attention for the same. The so-called "distance-sense of the blind" is a proof of this. It consists in the ability to perceive from some distance, without direct contact, a resisting object, as, for example, a neighboring wall. Now, it can be experimentally demonstrated that this distance-sense is made up of two factors: a very weak tactual stimulation of the forehead by the atmospheric resistance, and a change in the sound of the step. The latter acts as a signal to concentrate the attention enough so that the weak tactual stimulations can be perceived. The "distance-sense" disappears, accordingly, when the tactual

stimulations are prevented by binding a cloth around the forehead or when the steps are rendered inaudible.

12. Besides our ideas of the position and movements of the various parts of our body, we have also an idea of the *position and movement of our whole body.* The former can never have anything but a relative significance; it is only when considered in connection with the latter that they become absolute. The organ of orientation for this general idea is the head. We always have a definite idea of the position of the head; the other organs are localized in our ideas, generally, indeed, very indefinitely, with reference to it, according to the particular complexes of inner and outer tactual sensations in each case. The specific organ of orientation in the head is the system of semicircular canals, to which are added as secondary aids the inner and outer tactual sensations resulting from the action of the muscles of the head. The function of these canals as an organ of orientation can be most easily understood by assuming that inner tactual sensations with especially marked differences in local signs, arise in them through the influence of the changing pressure of the fluid medium, which fills them. It is highly probable that *dizziness,* which comes from rapid rotation of the head, is due to the sensations caused by the violent movements of this fluid. This is in accord with the observations that partial derangements of the canals bring about constant illusions in localization, and complete derangement of the same is followed by an almost total suspension of the ability to localize.

12a. The antagonistic theories in regard to the psychical formation of spacial ideas, are generally called *nativism* and *empirism.* The *nativistic* theory seeks to derive localization in space from connate properties of the sense-organs and sense-centers, while the *empiristic* theory seeks to derive it from the influences of experience. This discrimination does not give proper expression to the actual opposition that exists, for the assumption of connate spacial ideas may be attacked without affirming that these ideas arise through experience. This is the case when, as above, space-perceptions are regarded as products of psychical fusions due both to the physiological properties of the organs of sense and of movement, and to the general laws for the rise of psychical compounds. Such processes of fusion and the arrangements of sense-impressions based upon them, are everywhere the substrata of our experience, but for this very reason it is inadmissible to call them "experience" itself. It is much more proper to point out the opposition that really exists, as that of *nativistic* and *genetic,* theories. It is to be noted that the widespread nativistic theories contain empirical elements, while, on the other hand, empirical theories contain nativistic elements, so that the difference is sometimes very small. Supporters of the nativistic view assume that the arrangement of impressions in space corresponds directly to the arrangement of sensitive points in the skin and retina. The special way in which the projection outward is effected, especially the ideas of the distance and magnitude of objects, and the reference of a plurality of specially separated impressions to a single object, are all regarded as dependent upon "attention", "will", or even "experience". Supporters of the empirical theory, on the other hand, generally presuppose space as given in some way or other, and then interpret each single idea as a localization in this space due to some empirical motive. In the theory of spacial ideas from sight, tactual space is generally regarded as this originally given space; in the theory of tactual ideas, original spacial qualities have sometimes been attributed to motor sensations. Thus, in the actual concrete theories empirism and nativism are very ill defined concepts. They agree in the use of the complex concepts of popular psychology, such as "'attentions", "will", and "experience", without any examination or analysis. In this respect they are different from the *genetic* theory, which seeks to show the elementary processes from which the ideas rise, by means of a psychological analysis of the ideas. In spite of their weaknesses, the nativistist and empiristic theories have served to set the psychological problem that exists here, clearly before us and to bring to light a great number of facts for its solution.

B. SPACIAL SIGHT-IDEAS

13. The general properties of the touch-sense are repeated in the visual sense, but in a more highly organized form. Corresponding to the sensory surface of the outer skin, we have here the retina with its rods and cones arranged in rows and forming an extraordinarily fine mosaic of sensitive points. Corresponding to the movements of the tactual organs, we have the movements of both eyes in fixating objects and following their bounding lines. Still, while tactual impressions are perceived only through immediate contact with the objects, the refractive media in front of the retina throw inverted, reduced images upon it. These images are so small that space is allowed for a large number of simultaneous impressions, and the ability of light to traverse space makes it possible for both neighboring and distant objects to yield impressions. Vision thus becomes a *distance-sense* in a much higher degree than hearing. Light can be perceived from incomparably greater distances than sound. Furthermore, only visual ideas are *directly* localized at different distances from the subject; for auditory ideas this localization is always indirect, through the aid of visual ideas of space.

14. With regard to its spacial attributes, every visual idea may be resolved into two factors: 1) the location of the single elements in relation to one another, and 2) their location in relation to the ideating subject. Even the idea of one single point of light, contains both these factors, for we must always represent a point in some spacial environment, and also in some direction and at some distance from ourselves. These factors can be separated only through deliberate abstraction, never in reality, for the relation of any point in space to its environment, regularly determines its relation to the ideating subject. As a result of this dependence, the analysis of visual ideas may better start with the location of the elements in relation to one another, and then take up later the location of the compound in regard to the subject.

a. The Location of the Elements of a Visual Idea in Relation to One Another.

15. In the apprehension of the reciprocal relations between elements of a visual idea, the attributes of the tactual sense are all repeated, only in a much more highly organized form, and with a few modifications significant for the visual ideas. Here, too,

we immediately connect with the simplest possible, approximately punctiform, impression the idea of its place in space; that is, we give it a certain definite position in relation to the parts of space about it. This localization is not effected, however, as in touch, by the direct reference of the impression to the corresponding point of the sense-organ itself, but we project it into a *field of vision,* which lies at some distance outside of the ideating subject. Here too we have a measure, as in the case of touch, for the accuracy of localization, in the distance at which two approximately punctiform impressions can be just distinguished as spacially different. The distance is not given in this case is a directly measurable linear extension on the sensory surface itself, but as the shortest perceptible interval between two points in the field of vision. The field of vision may be thought of as placed at any distance whatever from the subject, so that it is best to use as a measure for the fineness of localization, not a linear extension, but an *angle*, the angle formed by the intersection of the lines passing through the nodal point of the eye, from the points in the field of vision to the corresponding retinal points. This *angle of vision* remains constant so long as the size of the retinal image is unchanged, while the distance between the points in the field of vision increases in proportion to their distance from the subject. If an equivalent linear distance is sought in place of the angle of vision, it can be found in the diameter of the retinal image. This may be reckoned directly from the angle and the distance of the retina from the nodal point of the eye.

16. The measurements of the *keenness of localization with the eye*, made according to this principle, show that there is a great difference for different parts of the field of vision, just as was found for different regions of the tactual organs. Still, the distances that measure the smallest perceptible intervals here are all very much smaller. Then, again, there are many regions of finer discrimination scattered over the tactual organ, but only one region of finest discrimination in the field of vision. This is the middle, which corresponds to the center of the retina. From this region towards the periphery the fineness of localization diminishes very rapidly. The whole field of vision or the whole retinal surface, is, accordingly, analogous to a single tactual region, as, for example, that of the index finger, except that it very much surpasses the latter in fineness of localization, especially at the center, where two impressions at a distance corresponding to 60" - 90" in the angle of vision, are just distinguishable, while two degrees and a half toward the periphery, the smallest perceptible extension is 3' 30", and at eight degrees it increases to 1°.

In normal vision we turn the eye towards objects of which we wish to gain more accurate spacial ideas, in such a way that they lie in the middle of the field of vision, their images falling, accordingly, on the center of the retina. We speak of such objects as seen *directly,* of all others, which lie in the eccentric parts of the field of vision, as seen *indirectly.* The center of the region of direct vision is called the *point of regard,* or the *fixation-point.* The line that unites the center of the retina with that of the field of vision is known as the line *of regard.*

If we reckon the distance on the retina that corresponds to the smallest angle of vision at which two points in the center of the field of vision may be perceived as separate, it will be found to be .004 to .006 mm. This distance is equal to the diameter of a retinal cone, and since the cones are so close together in the center of the retina that they are in direct contact, it may be concluded with probability that two impressions must fall upon at least two different retinal elements if they are to be perceived as separate in space. This view is supported by the fact that in the peripheral regions of the retina the rods and cones, which are the two forms of elements sensitive to light, are really separated by greater intervals. It may, then, be assumed that the *keenness of vision,* or the ability to distinguish two distinct points in the field of vision, is directly dependent on the proximity of the retinal elements to one another, for two impressions can be distinguished as specially different only when they act upon different elements.

16a. Because of this interrelation between the keenness of vision and the arrangement of retinal elements, it has often been concluded that every such element has from the first the property of localizing any stimulus that acts upon it, in that position in space which corresponds to its own projection in the field of vision. In this way the attempt has been made to reduce the property of the visual sense by virtue of which it represents its objects in an external field of vision at some distance from the subject, to a condate energy of the retinal elements or of their central connections in the visual center in the brain. There are certain pathological disturbances of vision that seem at first sight to confirm this assumption. When some region of the retina is pushed out of place as a result of inflammation underneath, certain distortions in the images, the so-called *metamor-phopsia,* arise. The extent and direction of these distortions can be fully explained when it is assumed that the displaced retinal elements continue to localize their impressions as they did when in their normal positions. But it is obvious that these distortions of the images when, as in most cases, they appear as continually changing phenomena during the gradual formation and disappearance of the excretion, furnish us with no more evidence for a connate energy of localization in the retina than does the readily observed fact that distorted images of objects are seen when one looks through prismatic glasses. If, on the other hand, a stationary condition is gradually reached, the metamorphopsia disappear, and that, too, not only in cases where it may be assumed that the retinal elements return to their original position, but even in those cases where such a return is entirely improbable on account of the extent of the affection. In cases like the latter, the development of a new connection between the single retinal elements and their corresponding points in the field of vision, must be assumed. This conclusion is supported by observations made with normal eyes on the gradual adaptation to distorted images which are produced by external optical appliances. If a pair of prismatic glasses are worn before the eyes, marked and disturbing distortions of the images are the regular results. The straight bounding lines appear bent and the forms of the objects are thus distorted. These disturbances gradually disappear entirely if the glasses are worn some time. When the glasses are removed the distortions may appear in the opposite direction. All these phenomena can be understood if we presuppose that the spacial localizations of vision are not original, but *acquired.*

17. Besides the retinal sensations there are other psychical elements that always take part in the reciprocal spacial arrangement of light-impressions. The physiological

properties of the dye point a *priori* to the sensations that accompany *ocular movements*, as such elements. These movements obviously play the same part in the estimation of distances in the field of vision as the tactual movements do in the estimation of tactual impressions. The grosser conditions of touch are, however, here reproduced in a much more delicate and highly developed form. The eye can be turned in all directions about its center of rotation, which is fixed in its relation to the head, by means of a most admirably arranged system of six muscles. It is thus well suited to following continuously the bounding lines of objects or to passing each time in the shortest line from a given fixation-point to another. The movements in the directions which correspond to the position of the objects most frequently and closely observed, namely, downward and inward movements, are favored above the others by the arrangement of the muscles. Furthermore, the movements of the two eyes are so adapted to one another through the synergy of their innervation, that normally the two lines of regard are always turned upon the same fixation point. In this way a cooperation of the two eyes is made possible which not only permit a more perfect apprehension of the position of objects in relation to one another, but, more especially, furnishes the most essential means for the determination of the spacial relation of objects to the subject (24 seq).

18. The phenomena of vision teach that the idea of the *relative distance* of two points from each other is dependent on the motor energy of the eye employed in passing through this distance, just as the discrimination of two distinct points in the field of vision depends on the arrangement of the retinal elements. The motor energy becomes a component of the idea through its connection with a *sensation* which can be perceived, especially in extensive movements and by comparing ocular movements in various directions. Thus, for example, the upward movement of the eyes is clearly accompanied by more intense sensations than the downward movements; and the same is true of outward movements of one eye as compared with its inward movements.

The influence of sensations of movement on the localization are most apparent in the cases of disturbance arising from partial paralysis of single ocular muscles. These disturbances correspond exactly to the changes in the amount of energy required to move the eye. The general principle of such disorders is that the distance between two points seems greater when they lie in the direction of the more difficult movement. The more difficult movement has a correspondingly more intense motor sensation, which under normal conditions accompanies a more extensive movement. As a result, the distance passed through appears greater. Furthermore, the same illusion may appear for distances that lie in the same direction, bait have not been actually passed through, for the standard found during a, movement determines the motor impulse in the eye even when it is not moved.

19. Similar differences in the estimation of distances can be demonstrated for the normal eye. Although the ocular muscles are so arranged that their movements in various directions require about the same amount of exertion, still, this is not exactly so. The reasons are apparently closely connected with the adaptation of the eye to its functions. The neighboring objects of our immediate environment, on which the lines of regard must be converged, are the ones most often looked it. For this reason, the

muscles of the eye have so adapted themselves that the movements for the convergence of the lines of regard are the easiest, particularly those directed downwards as compared with other possible movements of convergence. This general facilitation of convergence has been acquired by the addition of special auxiliary and compensatory muscles (superior and inferior oblique) to the muscles that move the eye upwards and downwards (superior and inferior recti). As a result of the greater complexity of muscular activity thus necessary for the upward and downward movements of the eyes, the exertion is greater in these directions than towards the two sides, where only the internal and external recti act. The relative case of downward movements of convergence shows itself partly in the differences in the intensity of sensations accompanying the movements, as already remarked, and partly in the fact that downward convergence is involuntarily too great and upward too small. There are certain *constant optical illusions depending on the direction of the object in the field of vision,* which correspond to these differences in the motor mechanism. They are of two kinds: illusions of *direction,* and those of *magnitude.*

Both eyes are subject to an illusion as to the *Direction of vertical lines in the field of vision.* Such a line whose upper end is inclined 1° - 3° outward, appears vertical, and one really vertical, seems inclined inward. Since the illusion is in opposite directions for the two eyes, it disappears in binocular vision. it can obviously be explained by the fact just noted, that the downward movements of the eyes are connected with an involuntary increase, and the upward movements with a decrease, in the convergence. This deflection of the movement from the vertical is not noticed, but we refer it to the object as a deflection in the opposite direction. An equally regular *illusion in magnitude* appears when we compare distances extending in different directions in the field of vision. This, too, is very probably to be referred to the asymmetry in the arrangement of the muscles which arises from the adaptation of the eye to the ordinary position of objects in space. A vertical straight line is judged on the average 1/6 too long as compared with an equal horizontal line. A square, accordingly, appears as a rectangle whose base is shorter than its sides, and a square drawn by the eye is always too low. This illusion is explained when we remember that, as a result of the highly developed tendency to convergence, the muscular activity for upward and downward movements is much more complex than for inward and outward movements. The consequence is the same as in the case of partially paralyzed eyes, distances in the direction of the more difficult movement appear greater.

19a. Besides this difference between vertical and horizontal distances, which is most noticeable because it is so large, there are less marked differences between upward and downward, as well as between outward and inward distances. The upper half of a vertical line is overestimated on the average by 1/16 of its length, and the outer half of a horizontal line by 1/40. The first way be due to a slight asymmetry in the arrangement of the upper and lower muscles, or it may be due to the involuntary convergence of the lines of regard in downward movement, or, finally, to a combination of both influences. The effect of convergence is due to the fact that it corresponds to an approach of the object, so that we are generally inclined to see the lower half of the line nearer. In accordance with certain conditions of association to be described later (§ 16,

9), when the angle of vision remains constant, whatever is judged as nearer is judged to be smaller, so that the lower half of a line seems shorter than the upper. This explanation by the perspective can not be applied to the greater illusion in the overestimation of vertical as compared with horizontal lines, for if it were applicable, the illusion would at most be about equal to that found in the comparison of the two halves of a vertical line, while in reality it is approximately *three times* as great. The fact that this greater illusion appears only when *straight* distances are compared, not in the case of objects bounded by curved lines, is also a proof against the explanation by perspective. A circle, for example, does not appear as an ellipse with a longer vertical axis, but as a real circle. The slight overestimation of the outer half of a horizontal line is also due most probably to the asymmetrical activity of the muscles, which arises from the relative ease of convergence-movements.

20. Added to these two illusions, which arise from the special structure of ocular muscles in their adaptation to the purposes of vision, there are certain other variable optical illusions that are due to certain attributes of all voluntary movements and have their analoga in the movements of the tactual organs. These illusions may also be divided into those of *direction,* and those of *magnitude.* The former follow the rule that acute angles are overestimated, obtuse underestimated, and that the direction of the intersecting lines varies correspondingly. For the illusions of magnitude we have the rule. forced or interrupted movements require more exertion than free and continuous ones. Any straight line that necessitates fixation is, accordingly, overestimated in comparison with an open distance marked off by two points, and a straight line interrupted by several dividing lines is overestimated in comparison with an uninterrupted line.

The tactual analogon of the illusion in angles is the tendency to overestimate small articular movements and to underestimate large ones. This comes under the general principle that a relatively greater expenditure of energy is required for a short movement than for a more extensive one, because it is more difficult to begin a movement than to continue it after it is already started. The tactual phenomenon analogous to the overestimation of interrupted lines, is that a distance estimated by a movement of one of the limbs always seems shorter when it is passed through in a single continuous movement than it does when the movement is several times interrupted. Here, too, the sensation corresponds to the expenditure of energy, which is, of course, greater for an interrupted than for a continuous movement. The overestimation of interrupted lines by the eye, takes place, as we can easily understand, only so long as no motives arise from the way in which the division is made, to hinder the movement of the eye over the interrupted line. Such a hindrance is present, for example, when the line is interrupted only once. This one point of division makes fixation necessary. If we compare such a line with a continuous one, we tend to estimate the first without any movement, with the point of division as a fixation-center, while the second is apprehended by a movement of the eye. As a result the continuous line seems longer than the interrupted line.

21. All these phenomena point to the immediate dependence of the apprehension of spacial directions and magnitudes on ocular movements. As further evidence pointing in the same direction, we have the negative fact that the arrangement of the retinal elements, especially their proximity to one another, normally has no influence on the ideas of direction and magnitude. This is most strikingly evident in the fact that the distance between two points appears the same whether served in direct or indirect vision. Two points that are clearly distinguished in direct vision, may become *one* in the eccentric parts of the field of vision, but so soon as they are distinguished at all, they will appear just as far apart in one region as in the other. This independence of the proximity of the retinal elements, in our perception of magnitude, holds even for a part of the retina that is not sensitive to light at all - for the *blind spot,* where the optic nerve comes into the eye. Objects whose images fall on the blind spot are not seen. The size of this spot is about 6°, and it is located 15° inward from the point of fixation. Images of considerable size, as, for example, that of a human face at a distance of six feet, may disappear entirely on it. Still, when points appear at the right and left or below and above this region, we localize them just as far from each other as we should in any other, uninterrupted part of the field of vision. The same fact is observed when some part of the retina becomes blind through pathological conditions. The resulting break in the field of vision shows itself only ill the fact that images falling on it are not seen, never through any changes in the localization of objects lying on opposite sides of the blind region.[9]

22. All these phenomena teach that the *keenness of vision* and the *apprehension of directions and distances in the field of vision,* are two different functions, which depend upon different conditions: the first depends *on the proximity of the retinal elements to one another,* the second on *ocular movements.* It follows directly that spacial ideas from sight can not be regarded as original and given immediately in the action of impressions of light with their spacial arrangement, any more than can the spacial ideas of touch. The spacial order is, here too, developed from the combination of certain sensational components which, taken separately, have no spacial attributes whatever. Other conditions also indicate that the sensational elements are related here in the same way as in the case of touch, and that the development of visual space under normal conditions runs entirely parallel to the development of space in congenital blindness, the only condition under which touch attains a similar independence. Retinal impressions correspond to impressions of contact, and ocular movements to touch-movements. Tactual impressions can gain spacial qualities only through the local coloring of the sensations connected with them - the local signs and in like manner - we must presuppose the same for retinal impressions.

22a. To be sure, a qualitative gradation of local signs on [128] the retina can not be demonstrated with the same evidentness as for the skin. Still, by the use of colors it can

---

[9] In this connection, we have the fact that the blind spot does not appear as a break in the field of vision, without sensational contents, but as a continuation of the general brightness and color of the whole field; for example, as white when we are looking at a white surface, as black when we look at a black one. This filling out of the blind spot is possible only through reproduced sensations, and is to be considered as one of the phenomena of association to be discussed later (§ 16).

be established in general that for greater distances from the retinal center the sensational quality gradually changes. The colors are not so saturated in indirect vision, and the color-tone also changes; for example, yellow appears orange. There is, indeed, in these properties of the retina no strict proof for the existence of pure local differences in the sensations, at least not in the fine gradations that must be assumed in the retinal center, for example. Still, they show that local differences in sensations do exist, and this seems to justify the assumption of such even beyond the limits of demonstration. This is all the more justifiable because here, where the gradations are much finer, the tendency to translate sensational differences directly into local differences, which has already been noticed in the case of touch, will be much more apt to destroy their specifically qualitative character. As a confirmation of this view we have the fact that the clearly demonstrable sensational differences at greater distances from the retinal center, can be observed only under favorable conditions of limited impressions, and disappears entirely when surfaces of uniform color are looked at. This disappearance of qualitative differences which are in themselves considerable, must be attributed in part at least to their relation to local differences. When, however, such relatively great differences disappear as a result of this relation, so that special methods are required for their demonstration, it can not be expected that very small differences will be demonstrable at all.

23. We assume, accordingly, qualitative local signs, which, judging from the data derived from the keenness of vision, are graded in the finest stages at the retinal center and more slowly in the eccentric parts. The formation of visual space may then be described as a combination of this system of local signs arranged in two dimensions, with a system of intensive sensations of movement. For any two local signs a and b there will be a corresponding sensation of movement   arising from the movement through the distance a b and serving as a measure of the same. A longer distance a c will have a more intense sensation of movement,   . Just as the point of finest discrimination on the finger is the center of reference, so in the same way the retinal center is such a point of reference for the eye. In fact, this is from the laws of ocular movements more obvious for the eye than it is for the tactual organ. Any luminous point in the field of vision is a stimulus for the center of ocular innervation, and tends to turn the line of regard reflexively upon itself. This reflex relation of eccentric stimuli to the retinal center is probably an essential condition for the development of the synergy of ocular movements mentioned above, and is, at the same time, an explanation of the great difficulty of observing objects in indirect vision. This difficulty is evidently due to the greater reflex impulse toward a point in indirect vision when the attention is concentrated upon it, than toward less favored points. As a result of the preeminent importance which the retinal center has for ocular movements, the point of fixation necessarily becomes the center of reference in the field of vision, and all distances in this field are brought under a unitary standard by being determined with reference to the fixation-point. The excitation of local sign is due to the action of external impressions, and both together cause the movement towards the retinal center. The whole process of visual space-arrangement is thus due to the fusion of three different sensational elements: first, the sensational qualities depending upon the character of the external stimulus, second the qualitative local signs depending on the places where the stimuli act, and third, the

intensive motor sensations determined by the relation of the stimulated points to the center of the retina. The latter elements may either accompany actual movements - this is the original case - or, when the eye remains at rest, are mere motor impulses of a particular intensity. Because of the regular connection between qualitative local signs and intensive sensations of movement, they may both together be regarded as a single system of *complex local signs.* The spacial localization of a simple visual impression, is a product of a complete fusion of the sensation caused by the external stimulus with the two interconnected elements belonging to this system of complex local signs. The arrangement of a number of simple impressions in space consists in the combination of a great number of such fusions, which are graded in quality and intensity according to the elements of the system of local signs. The predominating elements in these fusions are the sensations due to the external stimulation. In comparison with these, the elements of the system of local signs are so obscure, even in their original quality and intensity, that for the immediate apprehension of objects they are entirely lost except as spacial qualities.

Connected with this complex process of fusion, which determines the order of the elements in the field of vision, is still another. This latter process, which takes place in the formation of every spacial idea, arises from the relation of the object seen to the subject. We pass now to the consideration of this second process.

b. *The Location of Visual Ideas in Relation to the Ideating Subject.*

24. The simplest case of a relation between an impression and the subject, that can appear in a visual idea, is evidently that in which the impression is reduced to a single point. If a single point of light is presented in the field of vision, both lines of regard are turned upon it as a result of the reflex impulse exerted by the stimulus, in such a way that in both eyes the images fall upon the retinal centers. At the same time the organs of accommodation are adapted to the distance of the point. The point thus represented on the centers of both retinas is seen as *single* and as situated in a certain particular direction and at a certain particular distance from the ideating subject.

The subject is represented, as a rule, by a point that may be defined as the middle point of the straight line connecting the centers of rotation of the two eyes. We will call this the *point of orientation* for the field of vision, and the straight line drawn from it to the intersection of the two lines of regard, that is to the external fixation-point, the *line of orientation.* When a point in space is fixated, there is always a fairly exact idea of the *direction* of the line of orientation. This idea is produced by the inner tactual sensations connected with the position of the two eyes. Such sensations are very noticeable because of their intensity, when the eyes are rotated much out of the central position. They are just as perceptible for a single eye, so that localization in direction is as perfect in monocular as in binocular vision. In the former case, however, the line of orientation generally coincides with the line of regard. [10]

25. The idea of the *distance* of the objects from the subject, or of the *absolute length* of the line of orientation, is much more indefinite than that of its direction. We are always inclined to ideate this distance shorter than it really is, as may be shown by comparing it with a standard placed somewhere in the field of vision perpendicular to the line of orientation. In this way we find that the distance on the standard judged to be equal to the line of orientation, is always much shorter than the real length of this line. The difference between the two increases further away the point of fixation moves, that is, the longer the line of orientation becomes. The only sensational components that can produce this idea of distance, are those inner tactual sensations connected with the position of the two eyes, that arise particularly from the convergence of the lines of regard and give somewhat of a measure for the absolute extent of this convergence. In fact, it is possible to observe sensations when the convergence is changed; mainly from the inner angle of the eye when the degree of convergence is increased, from the outer, when it is decreased. The sum of all the sensations corresponding to a given position of convergence distinguishes it completely from all other positions.

26. It follows that an idea of a definite, absolute length of the line of orientation can be developed only through the influences of experience, where in addition to the sensational elements a great many associations also have a part. This explains why these ideas always remain indefinite and why they are sometimes aided, sometimes interfered with by other components of visual ideas, especially by the size of the retinal images of familiar objects. On the other hand, we have a relatively fine measure in the sensations of convergence for *differences* in the distances of objects seen, that is, for the *relative* changes which the length of the line of orientation undergoes when the fixation-point approaches or recedes. For positions in which the lines of regard are nearly parallel, changes in convergence may be perceived that correspond to an angle of vision of 60" or 70". When the convergence increases, this least perceptible change in convergence also increases considerably, but, in spite of that, the corresponding differences in the length of the line of orientation become smaller and smaller. Thus the purely intensive sensations which accompany movements of convergence, are translated directly into ideas of changes in the distance between the fixation-point and the point of orientation of the subject.

This translation of a certain particular sensational complex into an idea of distance, is not due to any connate energy, but to a particular psychical development, as is shown by a great number of experiences which point to such a development. Among these is the fact that the apprehension both of absolute distances and of differences in distance, is greatly improved by practice. Children are generally inclined to localize very distant

---

[10] The habit of seeing with two eyes results in exceptions to this rule. Often when one eye is closed, the line of orientation remains the same as in binocular vision and does not coincide with the line of regard. In such cases the closed eye usually makes the corresponding movements of convergence upon a common fixation-point with the open eye.

objects in the immediate neighborhood: they grasp at the moon, at the slater on the tower, etc. In the same way, it has been observed that the congenitally blind are, immediately after an operation, entirely unable to distinguish near and far.

27. It is of importance for the development of this discrimination between far and near, that under the natural conditions of vision not mere isolated points are presented, but *extended three-dimensional objects,* or at least a number of points at different depths, to which we assign relatively different distances their respective lines of orientation.

Let us consider first the simplest case, where *two* points c and b are presented, lying at different depths, and connected by a straight line. A change in the fixation from a to b is always accompanied by a change in convergence, and brings about, first, the passage through a continuous series of retinal local signs corresponding to the line ab, and, secondly, a sensation of movement, [alp], corresponding to the difference in convergence between these two points. This gives us here, too, the elements of a spacial fusion. The product of this fusion is, however, peculiar in kind; it differs in both its components, in the successive series of local signs and in the concomitant sensations of movement, from the fusions that arise when a line in the field of vision is passed over. In the latter case the changes in local signs and sensations of movement are *alike* for both eyes, while in changing the point of fixation from far to near or the reverse they are opposite in the two eyes. For when the right eye a rotation towards the left convergence gives it will produce a rotation towards the right in the left eye, and vice versa. The same must also hold for the movement of the retinal images: when the image of the point as it leaves the point of fixation, moves towards the right in the right eye, it moves towards the left in the left eye, and vice versa. The first takes place when the eyes turn from a nearer to a more distant point, the latter, when they move in the opposite direction. Such fusions arising from movements of convergence have, so far as their qualitative and intensive components are concerned, a composition analogous to that on which the arrangement of the elements in the field of vision with regard to one another depends; but the special way in which these elements are united is entirely different in the two cases.

28. Thus, the fusions between local signs and sensations of convergence form a system of *complex local signs* which is analogous to that deduced above, but still peculiar in its composition. This system, differing in composition from the system of local signs in the field of vision, is supplementary to the latter in that it adds to the reciprocal relation between the objective elements a relation between the ideating subject and these elements. The relation to the subject divides into two ideational elements, characterized by peculiar sensational elements: the *idea of direction* and that of *distance.* Both refer primarily to the point of orientation in the head of the ideating subject, and are then secondarily applied to the relations of external objects in regard to one another. Thus, we come to assign to two points which lie at different distances along the line of orientation a direction and distance in relation to each other. All such ideas of spacial distance referring to various positions along the line of orientation, when taken together, are called *ideas of depth,* or when they are also ideas of particular single objects *ideas of three dimensions.*

29. An idea of depth arising in the way described varies according to objective and subjective conditions. The determination of the absolute distance of an isolated point in the field of vision, is always very uncertain. Even, the determination of the relative distance between two points *a* and *b* lying at different depths is generally certain only under the condition assumed above, that they are connected by a line along which the points of fixation for the two eyes can move in changing the convergence from *a* to *b*. We may call such lines which connect different points in space with one another *lines of fixation*. The principle may then be formulated: points in space are apprehended in their true relations only when they are connected by lines of fixation, along which the point of fixation may move. This principle is explicable on the ground that the condition of a regularly connected change in the local signs of the retina and in the accompanying sensations of convergence, that is, the condition for the rise of ideas of depth as we found before, is obviously fulfilled only when impressions are presented which can arouse the appropriate local signs.

30. When the condition mentioned is not fulfilled, there arises either an imperfect and indefinite idea of the different relative distances of the two points from the subject, or else the two points seem to the equally distant - a phenomenon which can appear only when one of the points is rigidly fixated. Under the latter condition still another charge, always arises in the idea; only the fixated point is seen as single, the other as double. The same thing happens while looking at objects when they are not connected with the binocular fixation-point by means of lines of fixation. Double images that arise in this way are *uncrossed* - *i.e.,* the right belongs to the right eye, the left to the left eye - when the crossed fixated point is nearer than the observed object and crossed when the point is beyond the object.

Binocular localization in depth and binocular double images are, accordingly, phenomena directly interrelated; where the former is indefinite and imperfect, we have double images, and where, on the other hand, the latter are absent, the, localization in depth is definite and exact. The two phenomena stand in such a relation to the line of fixation that, when it is present, localization is aided and double images removed. Still, this rule is not without exception, for when a point is rigidly fixated with both eyes, double images arise easily in spite of any lines of fixation that may be present. This is explained by the necessary conditions, for both of depth as mentioned above. Just as the absence of lines of fixation results in the lack of the required succession of the local signs, so in a singular manner the sensations of conference connected with movement, are absent in rigid fixation.

c. Relations between the Location of the Elements in regard to one another and the Location in Regard to the Subject.

31. When the field of vision is thought of as merely a location of impressions in relation to one another, we represent it to ourselves is a surface, and call the single objects lying in this surface *ideas of two dimensions,* in contrast to the ideas of depth. But even an idea of two dimensions must always be related to the seeing subject in two ways. First, every point in the field of vision is seen ill a particular *direction* on the subjective line of

orientation mentioned above. Secondly, the whole field of vision is localized at some *distance* or other from the subject, though this distance may be very indefinite.

The location in a particular direction results in an *erect* ideational object corresponding to an inverted retinal image. This relation between the objective localization in direction and the retinal image is as necessary a result of ocular movements as the inversion of the image itself is a result of the optical properties of the eye. Our line of orientation in space is the *external* line of regard or, for binocular vision, the middle line resulting from the combined effects of movements of fixation. A direction upward on this line of orientation in external space corresponds to a direction downward in the space where the retinal image lies, behind the center of ocular rotation, and vice versa. It follows that the retinal image must be inverted if we are to see the object erect.

32. The location at some distance or other, which is also never absent, brings about the result that all the points of the field of vision seem to be arranged on the *surface of a concave hemisphere* whose center is the point of orientation, or, in monocular vision, the point of the eye's rotation. Now, small areas of a large curved surface appear plane, so that the two-dimensional ideas of single objects are as a rule plane; thus, for example, figures drawn upon t plane, as those of plane geometry. But as soon as some parts of the general field of vision separate from it in such a way that they are localized before or behind, that is in different planes, the idea of two dimensions gives place to one of three.

32 a. The fusions formed between qualitative local signs and sensations of convergence when we change from the fixation of a more distant point to a nearer, or the reverse, may be called *complex local signs of depth.* Such local signs form for every series of points lying before or behind the fixation-point, or for an extended body which is nothing but a series of such points, a regularly arranged system in which a stereometric form located at a particular distance is always unequivocally represented by a particular fusion. When one of two points lying at different distances is fixated, the other is characterized by the different position of its images in the two eyes, and by the correspondingly different direction of the complex local signs in the two cases. The same is true for connected series of points or extended bodies. When we look at a solid object, it throws images in the two eyes that are different from each other on account of the different relative position of the object with regard to the two eyes. We may designate the difference between the positions of certain point in the image in the two eyes as the *binocular parallax.* This parallax is zero for the point fixated and for those points which are equally distant on the line of orientation; for all other points it has some real positive or negative value accordingly, as they are more or less distant than the fixation-point. If we fixate solid objects with both eyes, only the point fixated, together with those points which are equidistant and in its neighborhood in the field of vision, will give rise to images corresponding in position in the two eyes. All points of the object located at different distances, give images varying in position and size. These differences in the images are just what produce the idea of the solidity of the object when the proper lines of fixation are present. For in the way above described, the angle of binocular parallax for the image of any point lying before or behind the point of

fixation and connected with the same by a line of fixation, furnishes, according to its direction and magnitude, a measure for the relative distance of this point in depth through the complex local signs connected with the angle of parallax. This angle of parallax for a given objective depth, decreases proportionally to the distance of the solid object, so that the impression of solidity diminishes, the further of the objects are, and when the distance is so great that all angles of parallax disappear, the body will appear flat, unless the associations to be discussed later (§ 16, 9) produce an idea of depth.

33. The influence of binocular vision on the idea of depth may be investigated experimentally by means of a *stereoscope*. This instrument consists of two prisms with their angles of refraction turned toward each other in such a way that it renders possible a binocular combination of two plain drawings which correspond to the retinal images from a three-dimensional object. The influence of the various conditions that underlie the formation of ideas of depths, may, in this way, be studied much better than by looking at actual three-dimensional objects, for here we may vary the conditions at will.

To give a concrete illustration, it is observed that complex stereoscopic pictures generally require several movements of convergence back and forth before a clear plastic idea arises. Furthermore, the effect of the parallax appears in looking at stereoscopic pictures whose parts are movable in respect to each other. Such movements are accompanied by changes in the relief which answer exactly to the corresponding changes in binocular parallax. This parallax is dependent on the distance of the two eyes from each other, so that ideas of depth can be produced even in the case of objects too distant in reality to give a plastic effect, by combining in the stereoscope pictures taken from positions much further apart than the two eyes are. This is done, for example, in making stereoscopic photographs of landscapes. The result is that these photographs when combined do not look like real landscapes, but like plastic models regarded from a short distance.

34. In monocular vision all the conditions are absent which are connected with movements of convergence, and with binocular differences in the retinal images, and which may be artificially reproduced with the stereoscope. Still, not all the influences are wanting even here to produce a localization in the third dimension, although this localization is more imperfect.

The direct influence of *movements of accommodation* have in comparison with other conditions a relatively small, perhaps entirely insignificant influence. Still, like movements of convergence, they too are accompanied by sensations which can be clearly perceived in the else of greater changes of accommodation from distant to neighboring points. For smaller changes in depth these sensations are very uncertain. As a result the movement of a point in the direction of the line of regard, when it is looked at with only one eye, is generally not clearly observed until a change in the size of the retinal image appears.

35. For the development of monocular ideas of depth the influences which the components of the so-called *perspective* exercise, are of the greatest importance. These are the relative magnitude of the of vision, the trend of limiting lines, the direction of shadows, the change in colors due to atmospheric absorption, etc. All these influences, which act in exactly the same way in monocular and binocular vision, depend on *associations of ideas,* and will, therefore, be treated in a later chapter (§ 16).

35a. We have in general the same opposing theories for the explanation of visual ideas as for tactual ideas. The empirical theory has sometimes committed the fallacy of limiting itself to optics and turning the real problem of space perception over to touch. In such cases it has tried to explain only how a localization of visual ideas can take place with the aid of experience, on the basis of already existing spacial ideas from touch. Such an interpretation is, however, not only self-contradictory, but it also conflicts with experience, which shows that normal persons with vision, visual space-perception determines tactual, not the reverse. The fact of general development, that touch is the more primitive sense, can not be applied to the development of the individual. The chief evidences in support of nativistic theories are, first, the metamorphopsia after dislocation of retinal elements and, secondly, the position of the line of orientation, which indicates united functioning of the two eyes from the first. It has been noted already that the metamorphopsia and other related phenomena prove the exact opposite as soon as the chances to which they are due become stationary. Furthermore, the fact that in long continued use of only one eye the line of orientation comes to coincide with the line of regard, proves that the position of this line is not given from the first, but that it has arisen under the influence of the conditions of vision. Still another fact against nativistic and in favor of the genetic theory is the development in the child of the synergy of ocular movements under the influence of external stimuli and the organization of space-perceptions which apparently accompanies it. Here as in many other respects the development of most animals is different. In the latter cases the reflex connections of retinal impressions with movements of the eyes and head function perfectly immediately after birth. (v. inf. § 9, 2).

The *genetic* theory has gained the ascendancy over older nativistic and empirical views primarily through the more thorough investigation of the phenomena of *binocular vision.* Nativism has difficulty with the question why we generally see objects single although they produce images in each of the two eyes. The effort is made to avoid the difficulty by assuming that two identical retinal points are connected with the same optic fiber which divides in the chiasma, and that in this way they represent in the sensorium only a single point. This doctrine of the "identity of the two retinas" was, however, untenable as soon as the actual conditions of binocular vision in three-dimensions began to be investigated. Especially the invention of the *stereoscope* thus brought with it a new era for the genetic theory of vision.

11. TEMPORAL IDEAS

1. All our ideas are at once spacial and temporal. Just as the conditions for the spacial arrangement of impressions belong originally only to the tactual and visual senses, and just as spacial relations are only secondarily carried over from these to all other sensations, so there are only *two* kinds of sensations, namely, the inner tactual sensations from movements and the auditory sensations, which are the primary sources of temporal ideas. Still, there is a characteristic difference between spacial and temporal ideas in the fact that in the first the two senses mentioned are the only ones which can develop an independent spacial order, while in the second the two most important kinds of sensation are merely those in which the conditions are most favorable for the rise of temporal ideas. These conditions are not entirely wanting, however, for any sensations. This indicates that the psychological bases of temporal ideas are *more general,* and that they are not determined by the special structures of particular sense-organs. It follows from this fact that even when we abstract from the ideas that enter into any series of psychical processes, and take account only of the subjective phenomena accompanying the ideas, such as feelings and emotions, we still ascribe to the affective processes thus isolated through abstraction exactly the same temporal attributes as to the ideas. In philosophy the conclusion has generally been drawn from this fact that time is a "universal form of perception", that is, there is absolutely no psychical content that does not have a position in time, though such content may exist without spacial attributes. This conclusion that time-perception is more universal, arising, as it does, from the greater universality of the conditions of such perception, is erroneous and is not confirmed by psychological observation.

In the same way that we carry over spacial attributes from the two senses that give us space-perception to other kinds of sensations, we also give them secondarily to feelings and affective processes through the sensations and ideas inseparably connected with them. It may with equal right be doubted whether affective processes in themselves, without their related ideas, would have temporal attributes, for among the conditions of a temporal order are certain attributes of the sensational elements of ideas. The real facts in the case are that our ideas and, therefore, since ideas enter into every psychical experience, all psychical contents are at once spacial and temporal. The spacial order arises from certain particular sensational elements: in normal cases where vision is present from visual, in blindness, from tactual impressions; while time-ideas can arise from all possible sensations.

2. Temporal compounds like spacial and in contrast to intensive ideas, are characterized by the definite, unchangeable order of their component elements. If this order is changed, the given compound becomes another, even though the quality of its components remains the same. In special compounds, however, this unchangeableness of the order refers only to the relation of the elements to one another, not to the relation of the elements to the ideating subject. In temporal compounds, on the other hand, when the relation of one element is changed with regard to other elements, it is at the

same time changed with regard to the ideating subject. There is no change of position in time analogous to that possible in the case of space-compounds.

2a. This property of the absolute, strictly speaking unchangeable, relation of every temporal compound and every time-element, however short, to the ideating subject, is what we call the *flow of time.* Every moment in time filled by any content whatever has, on account of this flow, a relation to the ideating subject that no other moment can be substituted for it. With space the case is just reversed: the very possibility of substituting any spacial element in its relation to the subject for any other element whatever, is what gives rise to the idea of *constancy,* or absolute duration, as we express it, by applying a time-idea to a space-idea. The idea of *absolute* duration, that is of time in which no change takes place, is strictly speaking impossible in time-perception itself. The relation to the subject must change continually. We speak of an impression as lasting, when its single periods in time are exactly alike so far as their *sensational contents* are concerned, so that they differ only *in their relation to the subject.* The concept of duration when applied to time is, therefore, a merely relative concept. One time-idea may be more lasting than another, but no time-idea can have absolute duration, for without the double relation of different sensations to one another and to the ideating subject, no such ideas at all could arise. Even an unusually long unchanging sensation can not be retained. We interrupt it continually with other sensational contents.

We may, however, separate the two temporal relations always united in actual experience, that of the elements to one another and to the ideating subject, since each is connected with certain particular attributes of time-ideas. In fact, this separation of the two relations found its expression in particular words for certain forms of occurrence in time even prior to an exact psychological analysis of time-ideas. If the relation of the elements to one another is alone attended to, without regard to their relation to the subject, *temporal modes* come to be discriminated, such, for example, as brief, long, regularly repeating, irregularly changing, etc. If, on the contrary, the relation of the subject is attended to and the objective forms of occurrence abstracted from, we have as the chief forms of this relation the temporal *stages* past, present, and future.

A. TEMPORAL TOUCH-IDEAS

3. The original development of temporal ideas belongs to touch. Tactual sensations, accordingly, furnish the general substratum for the rise of both the spacial and temporal arrangements of ideational elements. The spacial functions of touch, however, come from the *outer* tactual sensations, while the inner sensations which accompany movements are the primary contents of the earliest temporal ideas.

The *mechanical* properties of the limbs are important physiological bases for the rise of these ideas. The arms and legs can be moved in the shoulder-joints and hip-joints by their muscles, and are at the same time subject to the action of gravitation drawing them downward. As a result there are two kinds of movements possible for them. First, we have those which are continually regulated by voluntary activity of the muscles and may, therefore, be indefinitely varied and accommodated at every moment to the existing needs - we will call these the arhythmical movements. Secondly, we have those in which the voluntary energy of the muscles is operative only so far as it is required to set the limbs oscillating in their joints and to maintain this movement - *rhythmical* movements. We may neglect for our present consideration the arhythmical movements exhibited in the various uses of the limbs. Their temporal attributes are in all probability derived from the rhythmical movements, and only a very indefinite comparison of the duration of irregular movements is possible.

4. With rhythmical movements the case is different. Their significance for the psychological development of time-ideas is due to the same principle which gives them their importance as physiological organs, namely, the principle of *the isochronism of oscillations of like amplitude.* In walking, the regular oscillations of our legs in the hip-joints not only make the muscular energy expended less, but reduce to a minimum the continual voluntary control of the movements. Furthermore, in natural walking the arms are supplementary aids.

Their oscillation is not interrupted at every step like that of the legs by the placing of the foot on the ground, so that they furnish because of their continuity a means for the more uniform regulation of their movements.

Every suite period of oscillation in such a movement is made up of a continuous succession of sensations that are repeated in the following period in exactly the same order. The two limits of the period are marked by a complex of *outer* tactual sensations: the beginning by the impression accompanying the removal of the foot from the ground, the end by that accompanying its return to the ground. Between these there is a continuous series of weak inner tactual sensations from the joints and muscles. The beginning and end of this series of inner sensations coincide with the outer sensations and are more intense than those between them. They arise from the impulse of movement coming to the muscles and joints and from the sudden inhibition of the same, and serve also to mark off the periods.

Connected with this regular succession of sensations is a regular and exactly parallel series of *feelings.* If we consider a single period in a series of rhythmical movements, there is always at its beginning and end a feeling of *fulfilled* expectation. Between the two limits of the period, beginning with the first movement, is a gradually growing, feeling of *strained expectation,* which suddenly sinks at the last moment from its maximum to zero, to make place for the rapidly rising and sinking feeling of fulfillment. From this point on the same series is again repeated. Thus, the whole process of a rhythmical of a touch-environment consists, on its affective side, of two qualitatively antagonistic feelings. In their general character these feelings belong to the direction of straining. and relaxing feelings. One is a momentary feeling, that is, one that rises very rapidly to its maximum and then sinks with equal rapidity; the other is a feeling of long duration which gradually reaches a maximum and then suddenly disappears. As a result, the most intense affective processes are crowded together at the extremities of the periods, and are made all the more intense through the contrast between the feeling of satisfaction and the preceding feeling of expectation. Just in the same way that this sharply marked limit between the single periods has its sensational substratum in the strong outer and inner tactual impressions that arise at this instant, as above mentioned, so we have a complete correspondence between the gradual rise of the feeling of expectation and the continuous series of weaker inner tactual sensations accompanying the oscillatory movements of the limbs.

5. The simplest temporal ideas of touch are made up of the rhythmically arranged sensations that follow one another with perfect uniformity in the manner described, when like oscillatory movements are repeatedly carried out. But even in ordinary walking a slight tendency toward a somewhat greater complication arises; the beginning of the first of *two* successive periods is emphasized, both in the sensation and in the accompanying feeling, more than the beginning of the second. In this case the rhythm of movement begins to be *metrical.* In fact, such a regular succession of accented and unaccented ideas corresponds to the simplest measure, 2/8 time. It arises easily in ordinary walking because of the physiological superiority of the right side, and appears very regularly when several persons are walking together in *marching.* In the latter case even more than two periods may be united into one rhythmical unit. The same is true of the complicated rhythmical movements of the dance. But in such composite tactual rhythms the auditory temporal ideas have a decided influence.

B. TEMPORAL AUDITORY IDEAS

6. The attribute of the auditory sense which most of all adapts it to the more accurate apprehension of the temporal relations in external processes, is the exceedingly short persistence of its sensations after the external stimulation; so that any temporal succession of sounds is reproduced with almost perfect fidelity in the corresponding succession of sensations. In close connection with this we have certain psychological properties of temporal auditory ideas. In the first place, they differ from temporal ideas of touch in that often only the extremities of the single intervals that go to make up the total idea, are marked by sensations. In such a case the relations of such intervals to one another are estimated essentially by the apparently empty or heterogeneously filled intervals that lie between the limiting sensations.

This is especially noticeable in the case of *rhythmical* auditory ideas. There are in general *two* possible forms of such ideas: continuous or only rarely interrupted successions of relatively lasting sensations, and *discontinuous* successions of strokes, in which only the extremities of the rhythmical periods are marked by external sounds. For a discontinuous succession of entirely uniform sounds the temporal attributes of the ideas are in general more apparent than for lasting impressions, since in the former case the influences of the tonal qualities are entirely wanting. We may confine our consideration to discontinuous series, because the principles that apply here hold for continuous successions also. In fact, the rhythmical division in the latter case, as may be easily observed, is made by means of certain single accents which are either given in the external impression or arbitrarily applied to it.

7. A series of regular strokes made in this way as the simplest form of temporal auditory ideas, is distinguished from the simplest form of temporal touch-ideas, described above, mainly by the absence of all objective sensational content in the intervals. The external impressions here do nothing but divide the separate intervals from one another. Still, the intervals of such a series are not entirely empty, but are filled by subjective affective and sensational contents which correspond fully to those observed in tactual ideas. Most emphatic of all are the *affective contents* of the intervals. These feelings in their successive periods of gradually rising and suddenly satisfied expectation, are the same as in the course of a rhythmical tactual movement. Even the sensational substratum for these feeling is not entirely absent; it is merely more variable. Sometimes it is nothing but the sensations of tension of the tympanum in their various intensifies. Then again it is the accompanying sensations of tension from other organs, or finally other sensations of movement in cases where an involuntary rhythmical movement is connected with the auditory series. But on account of the changeable character and generally small intensity of these motor sensations, the affective processes in auditory ideas 'are very much more clearly perceptible.

It follows from the conditions described that the influence of the subjective elements on the character of time-ideas is the easiest to demonstrate. First of all, this shows itself in the effect which different rates of the sensations have on the formation of temporal ideas. It is found that there is a certain medium rate of about 0.2 sec. which is most favorable for the union of a number of successive auditory impressions. Now, it is easy to observe that this is the rate at which the above mentioned subjective sensations and feelings are most emphatic in their alternation. If the rate is made much slower, the strain of expectation is too great and passes into an unpleasurable feeling which becomes more and more unendurable. If, on the contrary, the rate is accelerated, the rapid alternation of feelings becomes fatiguing. Thus, in both directions limits are approached where the synthesis of the impressions into a rhythmical time-idea is no longer possible. The upper limit is about one second, the lower about 0.1 sec.

8. Then again, this influence of the course of our sensations and feelings upon our apprehension of temporal intervals, shows itself just as clearly ill the changes that our idea of such an interval undergoes when the conditions of its apprehension are varied without changing its objective length. Thus, it has been observed that in general a period divided into intervals is estimated as longer than one not so divided. We have here a phenomenon analogous to that observed in the illusion with interrupted lines. The overestimation is generally much greater for temporal intervals. This is obviously due to the fact that the oft repeated alternation in sensations and feelings in an interval of time have a much greater influence than the interruption of the movement through points of division in the case of the similar spacial illusion. Furthermore, if in a long series of regular beats single impressions are emphasized by their greater intensity, or

91

by some qualitative peculiarity, the uniform result is overestimation of the intervals preceding and following the emphasized impression, in comparison with the other intervals of the same series. If, however, a certain rhythm is produced successively with weak and then with strong beats, the rate appears slower in the first case than in the second.

These phenomena are also explicable from the influence of the sensational and affective changes. An impression distinguished from the rest, demands a change in the course of the sensations, and especially of the feelings, preceding its apprehension, for there must be a more intense strain of expectation and a, correspondingly stronger feeling of relief or satisfaction. The feeling of expectation lengthens the interval preceding the impression, the feeling of relief that following. The case is different when the whole series is made up at one time of weak impressions, and at another of strong ones. In order to perceive a weak impression we must concentrate our attention upon it snore. The sensations of tension and the accompanying feelings are, accordingly, more intense, as may be easily observed, for weaker beats than for stronger ones. Here too, then, the different intensifies of the subjective elements that give rise to them are reflected in the differences between temporal ideas. The effect is, therefore, not only lost, but even reversed, when we compare not weak with strong but strong with still stronger beats.

9. The tendency found in the case of rhythmical touch-ideas for at least *two* like periods to unite and form a regular metrical unit shows itself in auditory ideas also, only in a much more marked degree. In tactual movements, where the sensations that limit the single periods are under the influence of the will, this tendency to form a rhythmical series shows itself in the actual alternation of weaker and stronger impressions. With auditory sensations, on the other hand, where the single impressions can be dependent only on external conditions, and are, therefore, objectively exactly alike, this tendency may lead to the following characteristic illusion. In a series of beats which are exactly alike in intensity and are separated by equal periods of time, certain single beats, occurring at regular intervals, are always heard as stronger than the others. The time that most frequently arises when there is nothing to determine it, is the 2/8-time, that is, the regular alternation of arses and theses. A slight modification of this, the 3/8-time, where *two* unaccented follow one accented beat, is also very common. This tendency to mark time can be overcome only by an effort of the will, and then only for very fast or very slow rates, where, from the very nature of the series, the limits of rhythmical perception are nearly reached. For medium rates, which are especially favorable to the rise of rhythmical ideas, a suppression of this tendency for any length of time is hardly possible. If the effort is made to unite as many impressions as possible in a unitary time-idea, the phenomena become more complicated. We have accents of different degrees which alternate in regular succession with unaccented members of the series and thus, through the resulting divisions of the whole into groups, umber of impressions that may be comprehended in a single idea is considerably increased. The presence of two different grades of accent gives 3/4-time and 5/8-time, the presence of three grades gives 4 /4-time and 6/4-time, and as forms with three feet we have 9/8-time and 12/8-time. More than three grades of accentuation or, when the unaccented note is counted, more than four grades of intensity, are not to be found in either musical

or poetical rhythms, nor can we produce more by voluntarily formation of' rhythmical ideas. Obviously, these *three grades of accentuation* mark the limits of the *possible complexity* of temporal ideas, in a way analogous to that in which the maximal number of included beats (§15, 6) marks the limits of their *length.*

The phenomenon of subjective accentuation and its influence on the sensation of rhythms, shows clearly that temporal ideas, like spacial ideas, are not derived from objective impressions alone, but that there are connected with these, subjective elements, whose character determines the apprehension of the objective impressions. The primary cause of the accentuation of a particular beat is always to be found in the increased intensity of the preceding and concomitant feelings and sensations of movement. This increase in the intensity of the subjective elements is then carried over to the objective impression, and makes the latter also seem more intense. The strengthening of the subjective elements may be *voluntary,* through the increase of the muscular strain which produces sensations of movement, and in this way, finally results in a corresponding increase in the feelings of expectation; or this strengthening may take place *without volition,* when the effort to perceive a number of impressions together brings about an immediate articulation of the temporal idea through the corresponding subjective sensational and affective variations.

C. GENERAL CONDITIONS FOR TEMPORAL IDEAS

10. If we seek to account for the rise of temporal ideas on the basis of the phenomena just discussed, and of the regular combination of subjective sensational and affective elements with objective impressions, as it is there apparent; we must start with the fact that a sensation thought of by itself, can no more have temporal that it could have spacial attributes. Position in time can be possible only when single psychical elements enter into certain characteristic relations with other such elements. This condition of the union of a number of psychical elements holds for temporal ideas just as much as for those of space, but the kind of union is characteristic, and essentially different from that in space-ideas.

The members of a temporal series *ab c d e f,* can all be immediately presented as a single whole, when the series has reached just as well as if they were a series of points in space. In the latter case, however, they would, on account of original ocular reflexes, be arranged in relation to the point of fixation, and this fixation-point could, at different times, be any one of the impressions *a* to *f.* In time-ideas, on the other hand, it is always the *impression of the present moment* in relation to which all the rest are arranged in time. When a new impression becomes, in a similar manner, the present impression, even though its sensational contents axe exactly the same as that of the earlier, still, it will be apprehended as *subjectively* different, for though the affective state accompanying a sensation may, indeed, be related to the feelings of another moment, the two can never be identical. Suppose, for example, that following the series *a b c d e f,* there is a second series of impressions, $a' b' c' d' e' f'$ in which $a' = a$, $b' = b$, $c' = c$, etc., so far as their sensation elements are concerned. Let us represent the accompanying feelings by   and ′ ′ ′ ′ ′ ′Then   and ′, and ′, and   ′, etc., will be similar feelings, because the sensations are the same; but they will not be identical, because every affective element depends not, only upon the sensation with which it is immediately connected but also upon the state of the subject as by the totality of its experiences. The state of the subject is different for each of the members of the series $a' b' c' d'. . .$, from what it was for the corresponding member of the series $a\ b\ c\ d$ because when the impression a' arrives, a has been present, and so a' can be referred back to a, while no such thing was possible in the case of a. Analogous differences in the affective states show themselves in composite series when repeated. These states are never identical, however much the subjective conditions of the momentarily present feelings may agree, for every one of them has its characteristic relation to the totality of psychical processes. If we assume, for example, a succession of a number of similar series $a\ b\ c\ d$, $a' b' c' d'$, $a'' b'' c'' d''$, etc., in which a equals a' and a'', b equals b' and b'', etc., so far as their sensational contents are concerned, still, a'' differs from a' in its affective conditions, for a' can be referred back only to a, while a'' can be referred back to both a' and a. Besides this, it is true that other differences between impressions like in themselves always trise from some chance accompanying sensations which influence the affective state.

11. Since every element of a temporal idea is arranged in relation to the impression immediately present, as above remarked, it follows that this present impression will have one of the attributes of the *fixation-point* in spacial compounds. It will be *more clearly* and *distinctly* perceived than other elements of the same idea. But there is a great difference in the fact that this most distinct perception is not connected, as in the case of spacial ideas, with the physiological organization of the sense-organ, but is due entirely to the general attributes of the ideating subject, as expressed in the affective processes. The momentary feeling accompanying the immediately present impression is what helps to its clearest apprehension. We may, accordingly, call the part of a temporal idea which forms the immediate impression the *fixation-point of the idea* or in general, since it does not depend on external structure, as does the fixation-point of spacial ideas, we may call it figuratively the inner *fixation-point*. The inner fixation-point is, then, that part of a temporal idea which corresponds to the most clearly *ideated* and the immediately present impression. The impressions that lie outside this point of fixation, that is, impressions that have preceded the present, are directly perceived. They are arranged in a regular gradation of diminishing degrees of clearness, from the fixation-point. A unitary temporal idea is possible only so long as the degree of clearness for each of its elements has some positive value. When the clearness of any element sinks to zero, the idea divides into its components.

12. The inner fixation-point of the temporal senses differs essentially from the outer fixation-point of the spacial senses, in that its character is primarily determined, not by sensational, but by *affective elements*. Since these affective elements are continually changing, in consequence of the varying conditions of psychical life, the inner fixation-point is also always changing. This change of the inner fixation-point is called the

continuous flow of time. By continuous flow we mean to express the fact that no moment of time is like any other, and that no such moment can return. This fact is connected with the one-dimensional character of time, which is due to this very circumstance, that the inner fixation-point of temporal ideas is continually moving forward, so that a single point can never recur. The arrangement of time in one dimension, with reference always to a changing point of fixation, in which the subject represents itself, is what gives rise to the result that the elements of time-Ideas have a fixed relation, not only with respect to one another, but also with respect to the ideating subject.

13. If we try to give an account of the means for the formation of this reciprocally interdependent order of the parts of an idea, and of their determination in regard to the ideating subject, it is obvious that these means can be nothing but certain of the elements of the idea itself,, which, considered in themselves, have no temporal attributes, but gain such attributes through their union. We may call these elements *temporal signs,* after the analogy of local signs. The characteristic conditions for the development of temporal ideas indicate from the first that these temporal signs are, in the main, *affective elements.* In the course of any rhythmical series every impression is immediately characterized by the concomitant feeling of expectation, while the sensation is of influence only in so far as it arouses the feeling. This may be clearly perceived when a rhythmical series is suddenly interrupted. Furthermore, the only sensations that are never absent as components of all time-ideas are the *sensation of movement.* In the case of tactual ideas these sensations of movement belong to the immediate elements of the ideas themselves, in auditory and other compounds that are brought into the time form, they are always present as subjective accompanying phenomena. We may, accordingly, regard the feelings of expectation as the *qualitative,* the sensations of movement as the *intensive, temporal signs* of a temporal idea. The idea itself must then be looked upon as a fusion of the two kinds of temporal signs with each other and with the objective sensations arranged in the temporal form. Thus, the sensations of movement, as a series of intensive sensations, give a uniform measure for the arrangement of the objective sensations as characterized in quality by the concomitant feelings.

13 a. The sensations of movement play a similar part in the formation of both time-ideas and space-ideas. This like sensational substratum leads very naturally to a recognition of a relation between these two forms of perception, which finds its expression in the *geometrical* representation of time by a straight line. Still, there is an essential difference between the complex system of temporal signs and the systems of local signs in the fact that the former is based primarily, not on the qualitative attributes of sensations, connected with certain special external sense-organs, but on *feelings* which may come in exactly the same way from the most widely differing kinds of sensation, since they are not dependent on the objective content of these sensations, but on their subjective synthesis. These characteristics of time-ideas account for the universal significance that we attribute to them. This was what was improperly expressed in the Kantian principle, that time is a "form of the inner sense". This

expression is to be criticized on the ground of its erroneous presupposition of an inner sense.

Here again we have the same opposed *natativist,* and *genetic* theories on the psychological origin of time-ideas, as we had in the case of spacial-ideas. In this case, however, nativism has never developed a theory in any proper sense. It usually limits itself to the general assumption that time is a "connate form of perception", without attempting to give any account of the influence of the elements and conditions of temporal ideas which can be actually demonstrated. The genetic theories of older psychology, as, for example, that of Herbart, seek to deduce time-perception from ideational elements only. This is, however, pure speculation and loses sight of the conditions given in actual experience.

12. COMPOSITE FEELINGS

1. In the development of temporal ideas it appears clearly that the discrimination of sensational and affective components in immediate experience is purely a product of abstraction. For time-ideas the abstraction proves impossible, because, in this case, certain feelings play an essential part in the rise of the ideas. Time-ideas may, therefore, be called *ideas* only when the final results of the process, the arrangement of certain sensations in relation to one another and to the subject, are considered; when their real composition is looked into, they are complex products of sensations and feelings. They are thus to a certain extent transitional forms between ideas and those psychical compounds that are made up of affective elements, and are designated by the general name *affective processes.* These affective processes resemble time-ideas especially in the impossibility of an abstract separation of the affective from the sensational elements in the investigation of their rise. This is due to the fact that in the development of all kinds of affective processes, sensations and ideas are determining factors, just as feelings are among the essential factors of temporal ideas.

2. *Intensive affective combinations,* or *composite feelings,* must be the first affective processes discussed, because in them the characteristic attributes of a single compound are the products of a momentary state. The description of the feeling, therefore, requires only the exact comprehension of the momentary condition, not a combination of several processes occurring in time and proceeding from one another. In this respect, the composite feelings stand in the same relation to emotions and volitions, which always consist of affective processes extending through periods of time, as intensive ideas do to extensive. Intensive psychical compounds, in the broadest sense of the term, include, accordingly, intensive ideas and composite feelings. Extensive compounds include as special forms of *temporal* arrangements, besides the temporal ideas, also emotions and volitions.

3. Composite feelings, then, are intensive states of unitary character in which single simple affective components are to be perceived. We may distinguish in every such feeling *component feelings* and a *resultant feeling.* The last component feelings are always simple sense-feelings. Several of these may unite to form a partial resultant which enters into the whole as a compound component.

Every composite feeling may, accordingly, be divided, 1) into a *total feeling* made up of all its components, and 2) into single *partial feelings* which go to make up the total feeling. These partial feelings are in turn of different grades according as they are simple sense-feelings (partial feelings of the first order) or feelings which are themselves composite (partial feelings of the second or higher orders). Where we have partial feelings of higher orders, complicated combinations or *interlacings* of the component elements may take place. A partial feeling of lower order may, at the same time, enter into several partial feelings of higher order. Such interlacings may render the nature of the total feeling exceedingly complicated. The whole may sometimes change its character, even when its elements remain the same, according as one or the other of the possible combinations of partial feelings takes place.

3 a. Thus, the musical chord *c e g* has a corresponding total feeling of harmony whose last elements, or partial feelings of the first order, are the feelings corresponding to the single clangs *c, e,* and *g.* Between these two kinds of feeling stand, as partial feelings of the second order, the three feelings of harmony from the double clangs *c e, e g* and *c g.* The character of the total feeling may have four different shades according as one of these partial feelings of the second order predominates, or all are equally strong. The cause of the predominance of one of these complex partial feelings may be either the greater intensity of its sensational components, or the influence of preceding feelings. If, for example, *c e g* follows *c e g* the effect of *c♭ e g* will be intensified, while if *c e g* follows *c e a* the same will hold for *c g.* Similarly, a number of colors may have a different effect according as one or the other partial combination predominates. In the last case, however, because of the extensive arrangement of the impressions, the spacial proximity has an influence antagonistic to the variation in the manner of combination and, furthermore, the influence of the spacial form with all its accompanying conditions is an essentially complicating factor.

4. The structure of composite feelings is, thus, in general exceedingly complicated. Still, there are different degrees of development even here. The complex feelings arising from impressions of touch, smell, and taste are essentially simpler in character than those connected with auditory and visual ideas.

The total feeling connected with outer and inner tactual sensations is designated in particular as the *common feeling,* since it is regarded as the feeling in which our total state of sensible comfort or discomfort expresses itself. From this point of view, the two lowest chemical senses, those of *smell* and *taste,* must also be regarded as contributors to the sensational substratum of the common feeling, for the partial feelings that arise from these two senses unite with those from touch to form inseparable affective complexes. In single cases, to be sure, one or the other of these feelings may play such an important part that the others disappear entirely. Still, in the midst of all this change in its sensational substratum, the common feeling is always the immediate expression of our sensible comfort and discomfort, and is, therefore, of all our composite feelings most closely related to the simple sense-feelings. Auditory and visual sensations, on the other hand, contribute to the sensational substratum of the common feeling only in exceptional cases, especially when the intensity is unusually great.

4a. The combination of partial feelings to a composite feeling was first noticed in the case of the common feeling. The psychological laws of this combination were indeed misunderstood, and, as is usually the case in physiology, the feeling was not distinguished from its underlying sensations. Common feeling was, thus, sometimes defined as the "consciousness of our sensational state", or again as the "totality, or unanalyzed chaos of sensations" which come to us from all parts of our body. As a matter of fact, the common feeling consists of a number of partial feeling. But it is not the mere sum of these feelings; it is rather a resultant total feeling of unitary character. At the same time it is, however, a total feeling of the simplest possible composition, made up of partial feelings of the first order, that is, of single sense-feelings which generally do not unite to form partial feelings of the second or of higher orders. In the resultant feeling a single partial feeling is usually predominant. This is regularly the case when a very strong local sensation is accompanied by a feeling of pain. On the other hand, weaker sensations may determine the predominant affective tone through their relatively greater importance. This is especially frequent in the case of sensations of smell and taste, and also in the case of certain sensations connected with the regular functioning of the organs, such as the inner tactual sensations accompanying the movements of walking. Often the relatively greater importance of a single sensation is so slight that the predominating feeling can not be discovered except by directing our attention to our own subjective state. In such a case the concentration of the attention upon it can generally make any partial feeling whatever predominant.

5. The common feeling is the source of the distinction between *pleasurable* and *unpleasurable feelings.* This distinction is then carried over to the single simple feelings that compose it, and sometimes even to all feelings. Pleasurable and unpleasurable are expressions well adapted to the indication of the chief extremes between which the common feeling, as a total feeling corresponding to the sensible comfort or discomfort of the subject, may oscillate; though to be sure, this feeling may not infrequently lie for a longer or shorter period in an indifference-zone. In the same way, these expressions may be applied to the single constituents so far as they go to make up one of the total feelings. On the other hand, it is entirely unjustifiable to apply these names to all other feelings, or, as is sometimes done, to make their applicability a necessary factor in the general definition of feeling. Even for the common feeling, pleasurable and unpleasurable can only be used as general class-names which include a number of qualitatively different feelings. This variety among feelings of the same class results from the very great variations in the composition of the single total feelings that we have included under the general name common feeling.

6. The composite character mentioned is the reason why there are common feelings which can not, strictly speaking, be called pleasurable or unpleasurable, because they contain elements belonging to both classes, and under circumstances either the one kind or the other may predominate. Such feelings made up of partial feelings of opposite character and deriving their characteristics from this combination, may be called *contrast-feelings*. A simple form of such among the common feelings is that of *tickling*. It is made up of a weak pleasurable feeling accompanying a weak external tactual sensation, and of feelings connected with muscular sensations aroused by the strong reflex impulses from the tactual stimuli. These reflex impulses may spread more or less, and often cause inhibitions of respiration when they reach the diaphragm, so that the resultant feeling may vary greatly in single cases in intensity, scope, and composition.

7. The composite feelings from sight and hearing are commonly called *elementary aesthetic feelings*. This name includes all feelings that are connected with composite perceptions and are therefore themselves composite. As a special form of feelings belonging to this class defined by the broader meaning of the term '   , we have those which are the elements of aesthetic effects in the narrower sense. The term elementary does not apply in this case to the feelings themselves, for they are by no means simple, but it is merely intended to express the relative distinction between these and still more composite higher aesthetic feelings.

The perceptive, or elementary aesthetic, feelings of sight and hearing may serve as representatives of all the composite feelings that arise in the course of intellectual processes, such as the logical, moral, and higher aesthetical feelings. For the general psychological structure of these complex affective forms is exactly that of the simpler perceptive feelings, except that the former are always connected with feelings and emotions that arise from the whole interconnection of psychical processes. While the extremes between which the common feelings move are chiefly the affective qualities that we call pleasurable and unpleasurable in the sense of personal comfort and discomfort, the elementary aesthetic feelings belong to the same affective direction, but in the more objective sense of *agreeable* and *disagreeable,* feelings. These terms express the relation of the object to the ideating subject rather than any personal state. It is still more apparent here than in the caged of pleasurable and unpleasurable feelings, that each of these terms is not the name of a single feeling, but indicates a general direction, to which belong an endless variety of feelings with individual peculiarities for each single idea. In single cases, too, but more variably, the other affective directions, those of the arousing and subduing, of the straining and relaxing feelings, may show themselves.

8. If we neglect for the moment this general classification mentioned, according to which the single forms are brought under the chief affective directions, all perceptive feelings may be divided into the *two* classes of *intensive* and *extensive* feelings, according to the relations which exist between the sensational elements and determine the quality of the feelings. *By intensive* feelings we mean those that depend on the relation of the qualitative attributes of the sensational elements of the ideas, by *extensive*

feelings those that arise from the spacial and temporal arrangement of the elements. The expressions "intensive" and "extensive" do not refer to the character of the feelings themselves, for they are in reality always intensive, but to the *conditions for the rise* of these feelings.

Intensive and extensive feeling are, accordingly, not merely the subjective concomitants of the corresponding ideas, but, since every idea consists usually of elements that are qualitatively different and of some extensive arrangement of these impressions, the same idea may be at once the substratum of both intensive and extensive feelings. Thus, a visual object made up of different colored parts arouses an intensive feeling through the mutual relation of the colors and an extensive feeling through its form. A succession of clangs is connected with an intensive feeling which corresponds to the qualitative relation of the clangs, and with an extensive feeling coming from the rhythmical or arhythmical temporal succession of the same. In this way, both intensive and extensive feelings are always connected with visual and auditory ideas, but, of course, under certain conditions one form may push the other into the background. Thus, when we hear a clang for just an instant, the only feeling perceived is the intensive feeling. Or when, on the other band, a rhythmical series of indifferent sounds is heard, only the extensive feeling is noticeable. For the purpose of psychological analysis it is obviously of advantage to produce Conditions under which one particular affective form is present and others so far as possible excluded.

9. When *intensive feelings* are observed in this way, it appears that those accompanying the *combination of colors* follow the rule that a combination of two colors whose qualitative difference is a maximum, also gives a maximal agreeable feeling. Still, every particular color-combination has its specific affective character made up of the partial feelings from the single colors, and of the total feeling arising as a resultant, of the same. Then, too, as, in the case of simple color-feelings, the effect is complicated by chance associations and the complex feelings coming from them. Combinations of more than two colors have not been adequately investigated.

The feelings connected with *combination of clangs* are exceedingly numerous and various. They constitute the affective sphere in which we see most clearly the formation of partial feelings of different orders discussed above, together with their interlacings varying under special conditions. The investigation of the single feelings that thus arise is one of the problems of the psychological aesthetics of music.

10. *Extensive* feelings may be subdivided into spacial and temporal. Of these, the first, or the *feelings of form,* belong mainly to vision, and the second, or the *feelings of rhythm,* to hearing, while the beginning of the development of 'both are to be found in touch.

The *optical feeling of forms* shows itself first of all in the preference of regular to irregular forms and then in the preference among different regular forms of those which have certain simple proportions in their various parts. The most important of these proportions are those of symmetry, or 1 : 1, and, of the golden section, or $x+1:x=$

x:1 (the whole is to the greater part as the greater part is to the smaller). The fact that symmetry is generally preferred for the horizontal dimensions of figures and the golden section for the vertical, is probably due to associations, especially with organic forms, such as that of the human body. This preference for regularity and certain simple proportions can have no other interpretation than that the measurement of every single dimension is connected with a sensation of movement and an accompanying sense-feeling which enters as a partial feeling into the, total optical feeling of form. The total feeling of regular arrangement that arises at the sight of the whole form, is thus modified by the relation of the different sensations as well as of the partial feelings to one another. As secondary components, which also fuse with the total feeling, we may hive here too associations and their concomitant feelings.

The *feeling of rhythm* is entirely dependent on the conditions discussed in considering temporal ideas. The partial feelings are here the feelings of strained and fulfilled expectation, which in their regular alternation constitute the rhythmical time-ideas themselves. The way in which these partial feelings are united, however, and especially the predominance of special ones in the total feeling, is, even more than the momentary character of an intensive feeling, dependent on the relation in which the feeling present at a given instant stands to the preceding feelings. This is especially apparent in the great influence that every alteration in rhythm exercises on the accompanying feeling. For this reason as well as because of their general dependence on a particular temporal form of occurrence, the feelings of rhythm are the direct transitions to *emotions*. To be sure, an emotion may develop from any composite feeling, but in no other case is the condition for the rise of a feeling, as here, at the, same time a necessary condition for the rise of a certain degree of emotion. The emotion is, however, usually moderated in this case, through the regular succession of feelings (cf. § 13, 1, 7).

11. The immense variety of composite feelings and the equally great variety of their conditions, render any such comprehensive and at the same time unitary psychological theory as that which was possible for spacial and temporal ideas, entirely out of the question. Still, there are even here some common attributes, through which composite feelings may be brought under certain general psychological heads. There are two factors which go to make up every feeling: first, the relation of the combined partial feelings to one another, and second, their synthesis to a unitary total feeling. The first of these factors is more prominent in intensive, the second in extensive feelings. But in reality they are always united, and determine each other reciprocally. Thus, a figure which is all the time agreeable, may be more and more complex the more the relations of its parts accord with certain rules, and the same holds for a rhythm. On the other hand, the union to a single whole helps to emphasize the separate affective components. In all these respects combination of feelings show the closest resemblance to intensive ideas. The extensive arrangement of impressions on the contrary, especially the spacial arrangement, tends, much more to favor a relatively independent coexistence of several ideas.

12. The close intensive union of all the components of a feeling, even in the case of those feelings whose corresponding ideas are spacial or temporal, is connected with a

principle that holds for all affective processes, including those which we shall have to discuss later. This principle we will call that of the *unity of the affective state.* It may be formulated as follows: In a given moment only *one* total feeling is possible, or in other words, all the partial feelings, present at a given moment unite, in every case, to form a single total feeling. This principle of the unity of affective. states is obviously connected with the general relation between idea and feeling. For the "idea" deals with an immediate content of experience and the properties that belong to it, without regard to the subject; the "feeling" expresses the relation that invariably exists between this content and the subject.

13. EMOTIONS

1. Feelings, like all psychical phenomenal are never permanent states. In the psychological analysis of a composite feeling, therefore, we must always think of a momentary affective state as held constant. This is easier the more slowly and continuously the psychical processes occur, so that the word *feeling* has come to be used mainly for relatively slow processes and for those which in their regular form of occurrence never pass beyond a certain medium intensity, such as the feelings of rhythm. Where, on the other hand, a series of feelings succeeding one another in time unite to an interconnected process which is distinguished from preceding and following processes as an individual whole, and has in general a more intense effect on the subject than a single feeling, we call the unitary succession of feelings an *emotion.*

This very name indicates that it is not any specific subjective contents of experience which distinguish emotion from feeling, but rather the effect which comes from a special combination of particular affective contents. In this way it comes that there is no sharp line of demarcation between feeling and emotion. Every feeling of greater intensity passes into an emotion, and the separation between the two depends on a more or less arbitrary abstraction. In the case of feelings that have a certain particular form of occurrence [sic], that is *feelings of rhythm,* such an abstraction is strictly speaking impossible. The feeling of rhythm is distinguished at most by the small intensity of its moving effect on the subject, which is what gives "emotion" its name. Still, even this distinction is by no means fixed, and when the feelings produced by rhythmical impressions become somewhat more intense, as is usually the case, especially when the rhythm is connected with sensational contents that arouse the feelings greatly, they become in fact emotions. Feelings of rhythm are for this reason important aids both in music and poetry for portraying emotions and arousing them in the auditor.

The names of different emotions, like those of feelings, do not indicate single processes, but classes in which a large number of single affective processes are grouped on the ground of certain common characteristics. Emotions such as those of joy, hope, anxiety, care, and anger, are accompanied in any concrete case by peculiar ideational contents, while their affective elements also and even the way in which they occur may vary greatly from time to time. The more composite a, psychical processes is, the more variable will be its single concrete manifestations; a particular emotion, therefore, will be less apt to recur in exactly the same form than will a particular feeling. Every general name fore motions indicates, accordingly, certain *typical forms in which related affective processes occur.*

Not every interconnected series of affective processes is an emotion or can be classed as such under one of the typical forms discriminated by language. An emotion is a unitary whole which is distinguished from a composite feeling only through the two characteristics that it has a definite temporal course and that it exercises a more intense present and subsequent effect on the interconnection of psychical processes. The first characteristic arises from the fact that an emotion is a process of a higher order as compared with a single feeling, for it always includes a succession of several feelings. The second is closely connected with this first characteristic; it depends on the intensification of the effect produced by a summation of the feelings.

As a result of these characteristics emotions have in the

midst of all their variations in form a regularity in the manner of their occurrence. They always begin with a more or less intense *inceptive feeling* which is immediately characteristic in its quality and direction for the nature of the emotion, and is due either to an idea produced by an external impression (outer emotional stimulation) or to a psychical process arising from associative or aperceptive conditions (inner stimulation). After this inceptive feeling comes an *ideational process* accompanied by the corresponding feelings. This process shows characteristic differences in the cases of particular emotions both in the quality of the feelings and in the rapidity of the process. Finally, the emotion closes with a *terminal feeling* which continues even after the emotion has given place to a quiet affective state, and in which the emotion gradually fades away, unless it passes directly into the inceptive feeling of a new emotion. This last case occurs especially in feelings of the intermittent type (cf. inf. 13).

4. The intensification of the effect which may be observed in the course of an emotion, relates not merely to the psychical contents of the feelings that compose it, but to the *physical* concomitants as well. For single feelings these accompanying phenomena are limited to very slight changes in the innervation of the heart and respiratory organs, which can be demonstrated only by using exact graphic methods. With emotions the case is essentially different. As a result of the summation and alternation of successive affective stimuli there is here not only an intensification of the effect on heart, blood-vessels, and respiration, but the *external muscles* are always affected in an unmistakable manner. Movements of the oral muscles appear at first (mimetic movements), then movements of the arms and of the whole body (pantomimetic movements). In the case of stronger emotions there may be still more extensive disturbances of innervation, such as trembling, convulsive contractions of the diaphragm and of the facial muscles, and paralytic relaxation of the muscles.

Because of their symptomatical significance for the emotions, all these movements are called *expressive movements*. As a rule they are entirely involuntary, either reflexes following emotional excitations, or impulsive acts prompted by the affective components of the emotion. They may be modified, however, in the most various ways through voluntary intensification or inhibition of the movements or even through intentional production of the same, so that the whole series, of external reactions which we shall have to discuss under volitional acts, may take part in these expressive movements (§ 14). These different forms of movement may be entirely alike in external character and may pass into each other without sharp limitations on their psychical side, so that for the outside observer they are as a rule indistinguishable.

5. According to their symptomatical character, expressive movements may be divided into three classes. 1) *Purely* intensive symptoms; these are always expressive movements for more intense emotions, and consist of stronger movements for emotions of middle intensity, and of sudden inhibition and paralysis of movement for violent emotions. 2) *Qualitative expression of feelings;* these are mimetic movements, the most important of which are the reactions of the oral muscles, resembling the reflexes following sweet, sour, and bitter impressions of taste; the reaction for sweet corresponds to pleasurable emotions, those for sour and bitter to unpleasurable, while the other modifications of feeling, such as excitement and depression, strain and relief, are expressed by a tension of the muscles. 3) *Expression of ideas;* these are generally pantomimetic movements that either point to the object of the emotion (indicative gestures) or else describe the objects as well as the processes connected with them by the form of the movement (depicting gestures). Obviously these three classes of expressive movements correspond exactly to the psychical elements of emotions and their fundamental attributes: the first to their intensity, the second to the quality of the feelings, and the third to their ideational content. A concrete expressive movement may unite all three forms in itself. The third class, that of expressions of ideas, is of special psychological significance because of its genetic relations to *speech.*

6. The changes in *pulse* and *respiration* that accompany emotions are of three kinds. 1) They may consist of the immediate effects of the feelings that make up the emotions, as, for example, a lengthening of the pulse-curve and respiration-curve when the feelings are pleasurable, and a shortening of the same for unpleasurable feelings. This holds only for *relatively quiet* emotions where the single feelings have sufficient time to develop. When this is not the case, other phenomena appear which depend not merely on the quality of the feelings, but also, and that mainly, on the intensity of the innervations due to their summation. 2) Such summations may consist of *intensified* innervation, which arises from an *increase* in the excitation resulting from a summation when the succession of feelings is not too rapid. This increase shows itself in retarded and strengthened pulse-beats, since the intense excitation effects most the inhibitory nerves of the heart. Besides these there is usually an increased innervation of the mimetic and pantometic muscles. These are called *sthenic emotions.* 3) If the feelings are very violent or last an unusually long time in a single direction, the emotion brings about a more or less extended *paralysis* of the innervation of the heart and of the tension of the outer muscles. Under certain circumstances disturbances in the

innervation of special groups of muscles appear, especially those of the diaphragm and the sympathetic facial muscles. The first symptom of the paralysis of the regulative cardiac nerves is a marked acceleration of the pulse and a corresponding acceleration of the respiration, accompanied by a weakening of the same, and a relaxation of the tension of the external muscles to a degree equal to that in paralysis. These are the *asthenic emotions.* There is still another distinction, which is not important enough, however, to lead to the formation of an independent class of physical effects of emotions, since we have to do here only with modifications of the phenomena characteristic of sthenic and asthenic emotions. It is the distinction between *rapid and sluggish* emotions, based upon the greater or less *rapidity* with which the increase or inhibition of the innervation [*sic*] appears.

6a. Older psychology, following the method of Spinoza's famous doctrine of emotions, generally offered all kinds of logical reflections about emotions, for a theory of emotions or even for description of them. In recent times, on the other hand, the expressive movements and the other concomitants of emotion in the changes of innervation in pulse, respiratory organs, and blood-vessels, have attracted the most attention. Still, these phenomena, which are indeed valuable when rightly interpreted, are often used in a very wrong way as a means for the investigation of the psychological nature of affective processes. This has in turn led to a classification of emotions based entirely on their physical characteristics, and the strange theory has gained adherence that emotions are nothing but the results of expressive movements. The emotion of sorrow, for example, is regarded as made up entirely of the sensations that come from the mimetic of weeping. In a somewhat more moderate way the attempt has been made to use the expressive movements characteristics whose presence may be regarded as a mark to distinguish emotions from feelings. This is, however, unjustifiable since similar physical expressive phenomena appear even for the feelings, and the minor circumstance that these symptoms are in one case externally more or less clearly visible, evidently can not be decisive. The essential difference between emotion and feeling is *psychological.* The emotion is made up of a series of feelings united into a unitary whole. Expressive movements are the results, on the physical side, of the increase which the preceding parts of such a series have on those succeeding. It follows directly that the deciding characteristics for the classification of emotions must be *psychological.*

7. Though important constituents of emotions, the physical concomitants stand in no constant relation to the *psychical quality* of the same. This holds especially for the effects on pulse and respiration, but also for the pantomimetic expressive movements of stronger emotions. It may sometimes happen that emotions with very different, even opposite kinds of affective contents, may belong to the same class so far as the accompanying physical phenomena are concerned. Thus, for example, joy and anger may be in like manner sthenic emotions. Joy accompanied by surprise may, on the contrary, present the appearance, on its physical side, of an asthenic emotion. The *general* phenomena of innervation which give rise to the distinction between sthenic and asthenic, and rapid and sluggish emotions, do not show the character of affective contents of these emotions, but only the *formal* attributes of the intensity and rapidity of the feelings. This is clearly proved by the fact that differences in involuntary

innervation analogous to those which, accompany the different emotions, may be produced by a mere succession of indifferent impressions, as, for example, by the strokes of a metronome. It is observed in such a case that especially the *respiration* tends to adapt itself to the faster or slower rate of the strokes, becoming more rapid when the rapidity of the metronome increases. As a rule, too, certain phases of respiration coincide with particular strokes. To be sure, the hearing of such an indifferent rhythm is not unattended by emotion. When the rate changes, we observe at first a quiet, then a sthenic, and finally when the rapidity is greatest an asthenic emotion. Still the emotions in this case have to a certain extent a mere formal character; they exhibit a great indefiniteness in their contents. This indefiniteness disappears only when we think into them concrete emotions of like formal attributes. This is very easy, and is the condition of the great utility of rhythmical impressions for describing and producing emotions. All that is necessary to arouse an emotion in all its fullness is a mere hint of qualitative affective content, such as it is possible to give in music through the clangs of a musical composition.

7a. It follows from this relation of the physical effects to the psychical content of emotions, that the former can never be put in the place of the psychological observation of the emotions. They are general symptoms, but of such equivocal character that, though they are of great value when connected with introspection controlled by experimental methods, alone they have no value whatever. They are especially useful as cheeks for experimental introspection. The principle that the observation of psychical processes which present themselves in the natural course of life is entirely inadequate, holds especially for the emotions. In the first place, emotions come to the psychologist by chance, at moments when he is not in a condition to subject them to scientific analysis; and secondly, in the case of strong emotions whose causes are real we are least of all able to observe ourselves with exactness. This can be done much more successfully when we arouse in ourselves *voluntarily* a particular emotional state. In such a case, however, it is not possible to estimate how nearly the subjectively aroused emotion agrees in intensity and mode of occurrence [*sic*] with one of like character due to external circumstances. For this purpose the simultaneous investigation of the physical effects, especially of those most removed from the influence of the will of those on the pulse and respiration, furnishes a check for introspection. For when the psychological quality of emotions is alike, we may infer from their like physical effects that their formal attributes also agree.

8. Both in natural and in voluntarily aroused emotions, the physical concomitants have, besides their symptomatical significance, the important psychological attribute of *intensifying the emotion.* This attribute is due to the fact that the excitation or inhibition of certain particular groups of muscles is accompanied by inner tactual sensations which produce certain *sense-feelings.* These feelings unite with the other affective contents of the emotion and increase its intensity. From the heart, respiratory organs, and blood-vessels we have such feelings only for strong emotions, where they may indeed be very intense. On the other hand, even in moderate emotions the state of greater or less tension of the muscles exercises an influence on the affective state and thereby on the emotion.

9. The great number of factors that must be taken into consideration for the investigation of emotions renders a psychological analysis of the single forms impossible. This is all the more so because each of the numerous distinguishing names marks off a whole *class,* within which there is a great variety of special forms, including in turn an endless number of single cases of the most various modifications. All we can do is to take a general survey of the *fundamental form of emotions.* The general principles of division here employed must, of course, be *psychological,* that is, such as are derived from the immediate attributes of the emotions themselves, for the accompanying *physical* phenomena have only a symptomatical value and are even then, as noted above, equivocal in character.

Three such psychological principles of classification may be made the basis for the discrimination of emotions: 1) according to the *quality* of the feelings entering into the emotions, 2) according to the *intensity* of these feelings, 3) according to the *form of occurrence,* which is conditioned by the character and rate of the affective changes.

10. On the basis of *quality* we may distinguish certain fundamental emotional forms corresponding to the chief affective directions distinguished before. This gives us pleasurable and unpleasurable, exciting and depressing, straining and relaxing emotions. It must be noted, however, that because of their more composite character the emotions, are always, even more than the feelings, *mixed* forms. Generally, only a *single* affective direction can be called the *primary* tendency for a particular emotion. There are affective elements belonging to other directions, that enter in as secondary elements. Their secondary character usually appears in the fact that under different conditions various sub-forms of the primary emotion may arise. Thus, for example, joy is primarily a pleasurable emotion. Ordinarily it is also exciting, since it intensifies the feelings, but when the feelings are too strong, it becomes a depressing emotion. Sorrow is an unpleasurable emotion, generally of a depressing character; when the intensity of the feelings becomes somewhat greater, however, it may become exciting, and when the intensity becomes maximal, it passes again into very marked depression. Anger is much more emphatically exciting and unpleasant in its predominant characteristics, but when the intensity of the feelings becomes greater, as when it develops into rage, it may become depressing. Thus, exciting and depressing tendencies are always mere secondary qualities connected with pleasurable and unpleasurable emotions. Feelings of strain and relaxation, on the contrary, may more frequently be the chief, or at least the primary components of emotions. Thus, in expectation, the feeling of strain peculiar to this state is the primary element of the emotion. When the feeling develops into an emotion, it may easily be associated with unpleasurable feelings which are, according to circumstances either exciting or depressing. In the case of rhythmical impressions or movements there arise from alternation of feelings of strain with those of relaxation pleasurable emotions which may be either exciting or depressing according to the character of the rhythm. When they are depressing we may even have unpleasurable feelings intermingled with them, or they may all be of this kind, especially when other affective elements cooperate, for example feelings of clang or harmony.

11. Language has paid the most attention in its development of names for emotions to the *qualitative* side of feelings, and among these qualities particularly to pleasurable and unpleasurable. These names may be divided into *three* classes. First we have those of emotions that are *subjectively* distinguished, chiefly through the nature of the affective state itself, such as joy and sorrow and, as subforms of sorrow in which either depressing, straining, or relaxing tendencies of the feeling are also exhibited, sadness, care, grief, and fright. Secondly, there are names of *objective* emotions referring to some external object, such as delight and displeasure and, as subforms of the latter in which, as above, various tendencies unite, annoyance, resentment, anger, and rage. Thirdly, we have names of objective emotions that refer rather to outer events not expected until the future, such as hope and fear and, as modifications of the latter, worry and anxiety. They are combinations of feelings of strain with pleasurable and unpleasurable feelings and, in different ways, with exciting and depressing tendencies as well.

Obviously language has produced a much greater variety of names for unpleasurable emotions than for pleasurable. In fact, observation renders it probable that unpleasurable emotions exhibit a greater variety of typical forms of occurrence. and that their different forms are really more, numerous.

12. On the basis of the *intensity* of the feelings we may distinguish *weak* and *strong* emotions. These concepts, derived from the psychical properties of the feelings, do not coincide with those of sthenic and asthenic emotions, based upon the physical concomitants, for the relation of the psychological categories to the psycho-physical is dependent not only on the intensity of the feelings, but on their quality as well. Thus, weak and moderately strong pleasurable emotions are always sthenic, while, on the contrary, unplesurable emotions become asthenic after a longer duration, even when they are of a low degree of intensity, as, for example, care and anxiety. Finally, the strongest emotions, such as fright, worry, rage, and even excessive joy, are always asthenic. The discrimination of the psychical intensity of emotions is accordingly of subordinate significance, especially since emotions that agree in all other respects, may not only have different degrees of intensity at different times, but may on the same occasion vary from moment to moment. Then too since this variation from moment to moment is essentially determined by the sense-feelings that arise from the accompanying physical phenomena, in accordance with the principle of the intensification of emotions discussed above, it is obvious that the originally physiological antithesis of sthenic and asthenic often has a more decisive influence even on the psychological character of the emotion than the primary psychical intensity itself.

13. The third distinguishing characteristic of emotions the *form of occurrence,* is more important. Here we distinguish three classes. First, there are *sudden, irruptive* emotions, such as surprise, astonishment, disappointment, fright, and rage. They all reach their maximum very rapidly and then gradually sink to a quiet affective state. Secondly, we have *gradually arising* emotions, such as anxiety, doubt, care, mournfulness, expectation, and in many, cases joy, anger, worry. These rise to their maximum gradually and sink in the same way. As a third form and at the same time a modification

of the class just mentioned we have intermittent emotions, in which several, periods of rise and fall follow one another alternately. All emotions of long duration belong here. Thus, especially joy, anger, mournfulness, and the most various forms of gradually arising emotions, come in waves and often permit a distinction between periods of increasing and those of decreasing emotional intensity. The sudden, irruptive emotions, on the contrary, are seldom intermittent. This happens only in cases in which the emotion may also belong to the second class. Such emotions of a very changeable form of occurrence [sic] are, for example, joy and anger. They may sometimes be sudden and irruptive. In this case, to be sure, anger generally becomes rage. Or they may gradually rise and fall; they are then generally of the intermittent type. In their psycho-physical concomitants, the sudden irruptive emotions are all asthenic, those gradually arising may be either sthenic or asthenic.

13a. The form of occurrence [sic], then, however characteristic it may be in single cases, is just as little a fixed criterion for the Psychological classification of emotions as is the intensity of the feelings. Obviously such a classification can be based only on the quality of the affective contents, while intensity and form of occurrence may furnish the means of subdivision. The way in which these conditions are connected with one another and with the accompanying physical phenomena and through these with secondary sense-feelings, shows the emotions to be most highly composite psychical processes which are therefore in single cases exceedingly variable. A classification that is in any degree exhaustive must, therefore, subdivide such varying emotions as joy, anger, fear, and anxiety into their subforms, according to their modes of occurrence, the intensity of their component feelings, and finally according to their physical concomitants which are dependent on both the psychical factors mentioned. Thus, for example, we may distinguish a strong, a weak, and a variable form of anger, a sudden, a gradually arising, and an intermittent form of its occurrence, and finally a sthenic, asthenic, and a mixed form of its expressive movements. For the psychological explanation, an account of the causal interconnection, of the single forms in each particular case is much more important than this mere classification. In giving such an accounts we have in the case of every emotion to do with two factors, first, the quality and intensity of the component feelings, and second, the rapidity of the succession of these feelings. The first factor determines the general character of the emotion, the second its intensity in part and more especially its form of occurrence, while both together determine its physical accompaniments and the psycho-physical changes resulting from the sense-feelings connected with these accompanying phenomena. It is for this very reason that the physical concomitants are as a rule to be called *psycho-physical*. The expressions "psychological" and "psycho-physical" should not, however, be regarded as absolute opposites in this case, where we have to do merely with symptoms of emotion. We speak of psychological emotional phenomena when we mean those that do not show any immediately perceptible physical symptoms, even when such symptoms can be demonstrated with exact apparatus (as, for example, changes in the pulse and in respiration). On the other hand we speak of psycho-physical phenomena in the case of those which can be immediately recognized as two-sided.

14 VOLITIONAL PROCESSES

1. Every emotion, made up, as it is, of a series of interrelated affective processes having a unitary character, may terminate in one of two ways. It may give place to the ordinary variable and relatively unemotional course of feelings. Such affective processes that fade out without any special result, constitute the *emotions in the strict sense* as discussed in the last paragraph. The process may, in a second class of cases, pass into a *sudden* change in sensational and affective content, which brings the emotion to an instantaneous close; such changes in the sensational and affective state which are prepared for by an emotion and bring about its sudden end, are called *volitional acts.* The emotion itself together with its result is a *volitional process.*

A volitional process is thus related to an emotion as a process of a higher stage, in the same way that an emotion is related to a feeling. Volitional act is the name of only one part of the process, that part which distinguishes a volition from an emotion. The way to the development of volitions out of emotions is prepared by those emotions in connection with which external pantomimetic movements appear. These movements appear chiefly at the end of the process and generally hasten its completion; this is especially true of anger, but to some extent also of joy, care, etc. Still, in these mere emotions, the changes in the train of ideas which are the immediate causes of the momentary completion of the emotion in volitions and also the characteristic feelings attending these changes, are all wanting.

This close interconnection of volitional acts with pantomimetic movements necessarily leads us to look upon those volitions which end in certain bodily movements resulting from the preceding train of ideas and feelings, that is, those ending in *external* volitional acts, as the earliest stages in the development of volitions. The so-called *internal* volitional acts, on the other hand, or those which close simply with effects on ideas and feelings, appear in every case to be products of a more highly developed intelligence.

A volitional process that passes into an external act may be defined as an emotion which closes with a pantomimetic movement that has, in addition to the characteristics belonging to all such movements and due to the quality and intensity of the emotion, the special property of *producing an external effect which removes the emotion itself.* Such an effect is not possible for all emotions, but only for those which the very succession of component feelings produces feelings and ideas which are able to remove the preceding emotion. This is, of course, most commonly the case when the final result of the emotion is the direct opposite of the preceding feelings. The fundamental psychological condition for volitional acts is, therefore, the *contrast between feelings,* and the origin of the first volitions is most probably in all cases to be traced back to unpleasurable feelings that arouse external movements whose results are contrasted pleasurable feelings. The seizing of food to remove hunger, the struggle, against enemies to appease the feeling of revenge, and other, similar processes are original volitional processes of this kind. The emotions coming from sense-feelings, and the most wide spread social emotions, such as love, hate, anger, and revenge, are thus both for men and animals the common origin of will. A volition is distinguished in such cases from an emotion only by the fact that the former has added to its emotional components an external act that gives rise to feelings which, through contrast with the feelings contained in the emotion, bring the emotion itself to an end. The execution of the volitional act may then lead directly, as was originally always the case, or indirectly through an emotion of contrasted affective content, into the ordinary quiet flow of feelings.

3. The richer the ideational and affective contents of experience, the greater the variety of the emotions and the wider the sphere of volitions. There is no feeling or emotion that does not in some way prepare for a volitional act or at least have some part in such a preparation. All feelings, even those of a relatively indifferent character, contain in some degree an effort towards or away from some end. This effort may be very general and aimed merely at the maintenance or removal of the present affective state. While volitions appear as the most complex form of affective processes, presupposing all others -- that is, feelings and emotions - as their components, still, we must not overlook the fact that single feelings continually appear which do not unite to form emotions, and emotions appear which do not end in volitional acts. In the total interconnection of psychical processes, however, these three stages condition one another and form the related parts of a single process which is complete only when it becomes a volition. In this sense a feeling may be thought of as the beginning of a volition, or a volition may be thought of as a composite affective process, and an emotion may be regarded as an intermediate stage between the two.

4. The single feelings in an emotion that closes with a volitional act are usually far from being of equal importance. Certain ones among them, together with their related ideas, are prominent as those which are *most important* in preparing for the act. Those combinations of ideas and feelings which in our subjective apprehension of the volition are the immediate antecedents of the act, are called *motives* of volition. Every motive may be divided into an ideational and an affective component. The first we may call the *moving reason,* the second the *impelling force* of action. When a beast of prey seizes his victim, the moving reason is the sight of the same, the impelling force may be either the unpleasurable feeling of hunger or the race-hate aroused by the sight. The reason for a criminal murder may be the removal of an enemy, or some such idea, the impelling force the feeling of want, hate, revenge, or envy.

When the emotions are, of composite character, the reasons and impelling forces are generally mixed, often to so great an extent that it would be difficult for the author of the act himself to decide which was the leading motive. This is due to the fact that the impelling forces of a volitional act combine, just as the elements of a composite feeling do, to form a unitary whole in which all other impulses are subordinated under a single predominating one; the feelings of like direction strengthening and accelerating the effect, those of opposite direction weakening it. In the combinations of ideas and feelings which we call motives, the deciding importance in preparing for the act of will belongs to the feelings, that is, to the impelling forces, rather than to the ideas. This follows from the very fact that feelings are integral components of the volitional process itself, while, the ideas are of influence only indirectly, through their connections with the feelings. The assumption of a volition arising from pure intellectual considerations, of a decision opposed to the inclinations expressed in the feelings, is a psychological contradiction in itself. It rests upon the abstract concept of a transcendental will absolutely distinct from actual psychical volitions.

5. The combination of a number of motives, that is, of ideas and feelings which are distinguished in the composite train of emotions to which they belong, as those determining the discharge of the act, furnish the essential conditions for *the development of will,* and also for the discrimination of the *single forms of volitional action.*

The simplest case of volition is that in which a single feeling in an emotion of suitable constitution, together with its accompanying idea, becomes a motive and brings the processes to a close with its corresponding external movement. Such volitional processes determined by a *single* motive, may be called *simple volitions.* The movements in which they terminate are often designated *impulsive acts.* In popular parlance, however, this definition of impulse by the simplicity of the motive, is not sufficiently adhered to. Another element, namely, the character of the feeling that acts as impelling force, is here usually brought in. All acts that are determined by *sense-feelings,* especially common feelings, are generally called impulsive acts without regard to whether only a single motive or a plurality of motives is operative. This basis of discrimination is psychologically inappropriate and the complete separation of impulsive from volitional acts as a specifically distinct kind of psychical processes, which follows very naturally from it, is entirely unjustifiable.

By impulsive act, then. we mean a *simple* volitional act, that is, one resulting from a single motive, without reference to the position of this motive in the series of affective and ideational processes. Impulsive action, thus defined, must necessarily be the starting point for the development of all volitional acts, even though it may continue to appear along with the complex volitional acts. To be sure, the earliest impulsive acts are those which come from sense-feeling. In this sense most of the acts of animals are impulsive, but such impulsive acts appear continually in the case of man, partly as the results of simple sense-emotions, partly as the products of the habitual execution of certain volitional acts which were originally determined by complex motives.

6. When several feelings and ideas in the same emotion tend to produce external action, and when those components of an emotional train which have become motives tend at the same time towards different external ends, whether related or antagonistic, then there arises out of the simple act a *complex volitional process.* In order to distinguish this from the impulsive acts that precede it in the line of development, we call it a *voluntary act.*

Voluntary and impulsive acts have in common the characteristic of proceeding from single motives, or from complexes of motives that have fused together and operate as a single unequivocal impelling force. They differ in the fact that in voluntary acts the decisive motive has risen to do dominance from among a number of simultaneous and antagonistic motives. When a clearly perceptible strife between these antagonistic motives precedes the act, we call the volition by the particular name *selective act,* and the process preceding it a *choice.* The predominance of one over other simultaneous motives can be understood only when we presuppose such a strife in every case. But we perceive this strife now clearly, now obscurely, and now not at all. Only in the first case can we speak of a selective act in the proper sense. The distinction between voluntary and selective acts is by no means hard and fast. Still, in ordinary voluntary acts the psychical state is more like that in impulsive acts, while the difference between the latter and selective acts is clearly recognizable.

7. The psychical process immediately preceding the act, in which the final motive suddenly gains the ascendancy, is called in the case of voluntary acts *resolution,* in the case of selective acts *decision.* The first word indicates merely that action is to be carried out in accordance with some consciously adopted motive; the second implies that several courses of action have been presented as possible and that a choice has finally been made.

In contrast to the *first stages* of a volition, which can not be clearly distinguished from an ordinary emotional process, the *last stages* are absolutely characteristic. They are especially marked by accompanying *feelings* that never appear anywhere but in volitions, and must therefore be regarded as the specific elements peculiar to will. These feelings are first of all those of *resolution* and of *decision.* The latter differs from the former only in its greater intensity. They are both exciting and relaxing feelings, and may be united under various circumstances with pleasurable or unpleasurable factors. The relatively greater intensity of the feeling of decision is probably due to its contrast with the preceding feeling of *doubt* which attends the wavering between different motives. Its opposition to this doubt gives the feeling of relaxation a greater intensity. At the moment when the volitional act begins, the feelings of resolution and decision give place to the specific feeling of *activity,* which has its sensational substratum, in the case of external volitional acts, in the inner tactual sensation accompanying the movement. This feeling of activity is clearly exciting in its character, and is, according to the special motives of the volition, accompanied now by pleasurable, now by unpleasurable elements, which may in turn vary in the course of the, act and alternate with one another. As a total feeling, this feeling of activity is a rising and falling temporal process extending through the whole act and finally passing into the most various feelings, such as those of fulfillment, satisfaction, or disappointment, or into the feelings and emotions connected with the special result of the act. Taking the process as seen in voluntary and selective acts as *complete,* volitional acts, we must distinguish *compulsive acts* from them essentially by the absence of the antecedent feelings of resolution and decision. The feeling connected with the motive passes in the latter case directly into that of activity, and then into those which correspond to the effect of the act.

113

8. The transition from simple to complex volitional acts brings with it a number of other changes which are of great importance for the development of will. The first of these changes is to be found in the fact that the emotions which introduce the volitions lose their intensity more and more, as a result of the counteraction of different mutually inhibiting feelings, so that finally a volitional act may result from an apparently unemotional affective state. To be sure, emotion is never entirely wanting; in order that the motive which arises in an ordinary train of feelings may bring about a resolution or decision, it must always be connected with some degree of emotional excitement. This can, however, be so weak and transient that we overlook it. We do this the more easily the more we are inclined to unite a short emotion of this kind, attending merely the rise and action of the motive, with the resolution and execution in *the single* concept of a volitional act. This weakening of the emotions results mainly from the combinations of psychical processes which we call *intellectual* development and of which we shall treat more fully in the discussion of the interconnection of psychical compounds (§ 17). Intellectual processes can, indeed, never do away with emotions they are, on the contrary, in many cases the sources of new and characteristic emotions. A volition entirely without emotion, determined by a purely intellectual motive, is, as already remarked, a psychological impossibility. Still, intellectual development exercises beyond a doubt a moderating influence on emotions, particularly on those that prepare the way for volitional acts wherever intellectual motives enter into them. This may be due partly to the counteraction of the feelings which is generally present, partly to the slow development of intellectual motives, for in general emotions are the stronger the more rapidly their component feelings rise.

9. Connected with this moderation of the emotional components of volitions under the influence of intellectual motives is still another change. It consists in the fact that the act which closes the volition is not an external movement. The effect which removes the exciting emotion is itself a psychical process that does not show itself directly through any external symptom whatever. Such an effect which is imperceptible for objective observers is called an *internal volitional act.* The transition from external to internal volitional acts is so bound up with intellectual development that the very character of the intellectual processes themselves are to be explained to a great extent by the influence of volitions on the train of ideas (§ 15, 9). The act that closes the volition in such a case is some change in the train of ideas, which follows the preceding motives as the result of some resolution or decision. The feelings that accompany these acts of immediate preparation, and the feeling of activity connected with the change itself, agree entirely with those observed in the case of external volitional acts. Furthermore, action is followed by more or less marked feelings of satisfaction, of removal of preceding emotional and affective strain, so that obviously the only difference between these special volitions connected with the intellectual development and the earlier forms, is to be found in the fact that here the final effect of the volition does not show itself in an external bodily movement.

Still, we may have a bodily movement as the secondary result of an internal volitional act, when the resolution refers to an external act to be executed at some later time. In such a case the act itself always results from a special external volition whose decisive motives come from the preceding internal volition, but which we must consider as a new process distinct from the earlier. Thus, for example, the formation. of a resolution to execute an act in the future under certain expected conditions, is an internal volition, while the later, performance of the act is an external action different from the first, but requiring it as a necessary antecedent. It is evident that where an external volitional act arises from a decision after a conflict among the motives, we have a transition in which it is impossible to distinguish clearly between the two kinds of volition, namely that consisting in a single unitary process and that made up of *two* such processes, an internal and an external. In such a transitional form, if the decision is at all separated in time from the act itself, it may be regarded as an internal volitional act preparatory to the execution.

10. These two changes connected with the development of will, namely, the moderation of emotions and the rendering independent of internal volitions, are changes of aggressive order. In contrast with these there is a *third* process or one of *retrogradation.* When complex volitions with the same motive are often repeated, the conflict between the motives grows less intense; the opposing motives that were, overcome in earlier cases grow weaker and finally disappears

entirely. The complex act has then passed into a simple, *or impulsive act.* This retrogradation of complex volitional, processes into impulsive processes shows clearly the utter inappropriateness of the limitation of the concept "impulsive" to acts of will arising from sense-feelings. As a result of the gradual elimination of opposing motives, there are, intellectual, moral, and aesthetic, as well as simple sensuous, impulsive acts.

This retrogradation is but one step in a process that unites all the external acts of a living being, both the volitional acts and the automatic reflex movements. When the habituating practice of certain acts is carried further, the determining motives finally become, even in impulsive acts, weaker and more transient. The external stimulus originally aroused a strongly affective idea which operated as a motive, but now it causes the discharge of the act before it can be apprehended as an idea. In this way the impulsive movement finally becomes an automatic movement. The more often this automatic movement is repeated, the easier it, in turn, becomes, even when the stimulus is not sensed, as, for example, in deep sleep or during complete diversion of the attention. The movement now appears as a pure physiological reflex, and the volitional process has become a simple *reflex process.*

This gradual *reduction of volitional to mechanical processes,* which depends essentially on the elimination of all the elements between the physical beginning and end of the act, may take place either in the case of movements that were originally impulsive or in that of movements which have secondarily become such through the retrogradation of voluntary acts. It is not improbable that all the reflex movements of both animals and men originate in this way. As evidence for this we have, besides the reduction of volitional acts to pure mechanical processes through practice, as described above, also the *appropriate character of reflexes,* which point to the presence at some time of a purposive idea as motive. Furthermore, the circumstance that the movements of the lowest animals are all evidently simple volitional acts, not reflexes, tells for the same view, so that here too there is no justification for the assumption frequently made that acts of will have been developed from reflex movements. Finally, we can most easily explain from this point of view the facts mentioned before, that *expressive movements* may belong to any one of the forms possible in the scale of external acts. Obviously the simplest movements are impulsive acts, while many complicated pantomimetic movements probably came originally from voluntary acts which passed first into impulsive and then into reflex movements. Observed phenomena make it necessary to assume that the retrogradations that begin in the individual life are gradually carried further through the transmission of acquired dispositions, so that certain acts which were originally voluntary may appear ill later descendants from the first as impulsive or reflex movements (§ 19 and § 20).

10a. For reasons similar to those given in the case of emotions, the observation of volitional processes that come into experience by chance, is an inadequate and easily misleading method for establishing the actual facts in the case. Wherever internal or external volitional acts are performed in meeting either the theoretical or practical demands of life, our interest is too much taken up in the action itself to allow us at the same time to observe with exactness the psychical processes that are going on. In the theories of volition given by older psychologists - theories that very often cast their shadows in the science of to-day - we have a clear reflection of the undeveloped state of the methods of psychological observation. External acts of will are the only ones in the whole sphere of volitional processes that force themselves emphatically on the attention of the observer. As a result the tendency was to limit the concept will to external volitional acts, and thus not only to neglect entirely the whole sphere so important for the higher development of will, namely, internal volitional acts, but also to pay very little attention to the components of the volition that are antecedent to the external acts, or at most only to the more striking ideational components of the motive. It followed that the close genetic interconnection between impulsive and voluntary acts was not observed, and that ,the former were regarded as not belonging to will, but as closely related to reflexes. Will was thus limited to the voluntary and selective actions. Furthermore, the one-sided consideration of the ideational components of the motives led to a complete oversight of the development of volitional acts from emotions, and the singular idea found acceptance that volitional acts are not the products of antecedent motives and of psychical conditions which act upon these motives and bring one of them into the ascendancy, but that volition is a process *apart* from the motives and independent of them, a product of a metaphysical volitional faculty. This faculty was, on the ground of the limitation of the concept volition to voluntary acts, even defined as the choosing faculty of the mind, or as its faculty for preferring *one* from among the different motives that influence it. Thus, instead of deriving volition from the antecedent psychical conditions, the final result alone, the volitional act, was used to build up a general concept which was called *will* and this class-concept was treated in accordance with the faculty-theory as a first cause from which all concrete volitional acts arise. It was only a modification of this abstract theory when

116

Schopenhauer and, following him, many modern psychologists and philosophers declared that volition in itself is an "unconscious" occurrence which comes to consciousness only in its result, the volitional act. In this case, obviously, the inadequate observation of the volitional process preceding the act, has led to the assertion that no such process exists. Here, again, the whole variety of concrete volitional processes is supplanted by the concept of a *single* unconscious will, and the result for psychology is the same as before: in place of a comprehension of concrete psychical processes and their combination, an abstract concept is set up and then erroneously looked upon as a general cause.

Modern psychology and even experimental psychology is still to a great extent under the ban of this deep-rooted abstract doctrine of will. In denying from the first the possibility of explaining an act from the concrete psychical causality of the antecedent volitional process, it leaves as the only characteristic of an act of will the sum of the sensations that accompany the external act, and may immediately precede it as pale memo images in cases where the act has often been repeated. The physical excitations in the nervous system are regarded as the causes of the act. Here, then, the question of the causality is taken out of psychology and given over to physiology instead of to metaphysics, as in the theory discussed before. In reality, however, it is here too lost in metaphysics in attempting to cross to physiology. For physiology must, as an empirical science, abandon the attempt to give a complete causal explanation of a complex volitional act from its antecedents, not only for the present, but for all time, because this leads to the problem of an infinite succession. The only possible basis for such a theory is, therefore, the principle of materialistic metaphysics, that the so-called material processes are all that make up the reality of things and that psychical processes must accordingly be explained from material processes. But it is an indispensable principle of psychology as an empirical science, that it shall investigate the facts of psychical processes as they are presented in immediate experience, and that it shall not examine their interconnections from points of view that are entirely foreign to them. It is impossible to find out how a volition proceeds in any other way than by following it exactly as it is presented to us in immediate experience. Here, however, it is not presented as an abstract concept, but as a concrete single volition. Of this particular volition, too, we know nothing except what is immediately perceptible in the process. We can know nothing of an unconscious or, what amounts to the same thing for psychology, a material process which is not immediately perceived but merely assumed hypothetically on the basis of metaphysical presuppositions. Such metaphysical assumptions are obviously merely devices to cover up an incomplete or entirely wanting psychological observation. The psychologist who pays attention to only the termination of the whole volitional process, will very easily hit upon the thought that the immediate cause of volition is some unconscious immaterial or material agent.

11. The exact observation of volitional processes is, for the reasons given above, impossible in the case of volitional acts that come naturally in the course of life; the only way in which a thorough psychological investigation can be made, is, therefore, that of experimental observation. To be sure, we can not produce volitional acts of every kind at will, but we must limit ourselves to the observation of certain processes which can be

easily influenced through external means and which terminate in external acts. The experiments which serve this purpose are the so-called *reaction-experiments.* They may be described in their essentials as follows. A simple or complex volitional process is incited by an external sense-stimulus and then after the occurrence of certain psychical processes which serve in part as motives, the volition is brought to an end by a motor reaction.

Reaction-experiments have a second and more general significance besides that mentioned. They furnish means for the measurement of the *rate* of certain psychical and psycho-physical processes. In fact, such measurements are always made in these experiments. The primary significance of the experiments, however, consists in the fact that each one includes a volition and that it is therefore possible, in this way, by means of introspection to follow with exactness the succession of psychical processes in such a volition, and at the same time, by the deliberate variation of the conditions, to influence this succession in a systematic manner.

The simplest reaction-experiment that can be made is as follows. A short interval (2~3 see.) after a signal that serves to concentrate the attention, an external stimulus is allowed to act on some sense-organ. At the moment when the stimulus is perceived, a movement that has been determined upon and prepared before, as, for example, a movement of the hand, is executed. The psychological conditions in this experiment correspond essentially to those of a *simple* volition. The sensible impression serves as a simple motive, and this is to be followed invariably by a particular act. If now we measure objectively by means of either graphic or other chronometric apparatus, the interval that elapses between the action of the stimulus and the execution of the movement, it will be possible, by frequently repeated experiments of the same kind, to become thoroughly acquainted with the subjective processes that make up the whole reaction, while at the same time the results of the objective measurement will furnish a cheek for the constancy or possible variations in these subjective processes. This cheek is especially useful in those cases where some condition in the experiment and thereby the subjective course of the volition itself is intentionally modified.

Such a modification may, indeed, be introduced even in the simple form of the experiment just described, by varying the way in which the reactor *prepares,* before the appearance of the stimulus, for the execution of the act. When the expectation is directed toward the stimulus which is to serve as the motive,. the form of reaction known as *sensorial* results. When, on the other hand, the preparatory expectation is directed toward the act to be executed in response to the motive, we have the so-called *muscular* reaction. In the first case the ideational factor of the expectation is a pale memory-image of the familiar sense-impression. When the period of preparation is more extended, this image oscillates between alternating clearness and obscurity. The selective element is a feeling of expectation that oscillates in a similar manner and is connected with sensations of strain from the sense-organ to be affected, as, for example, with tension of the tympanum or of the ocular muscles of accommodation and movement. In the second case, on the other hand, where the reaction is muscular, we may observe during the period of preparatory expectation a pale, wavering memory-

118

image of the motor organ that is to react *(e. g.,* the hand) together with strong sensations of strain in the same, and a fairly continuous feeling of expectation connected with these sensations. Sensorial reaction-time is on the average 0.210-0.290 sec. (the shortest time is for sound, the longest for light), with a mean variation of 0.020 sec. for the single observations. Muscular reaction-time is 0.120-0.190 sec., with a mean variation of 0.010 see. The different values of the mean variation in the two cases are chiefly important as objective cheeks for the discrimination of these forms of reaction.[11]

12. By introducing special conditions we may make sensorial and muscular reactions the starting points for the study of the *development of volitions* in two different directions. Sensorial reactions furnish the means of passing from simple to complex volitions because we can in this case easily insert different psychical processes between the perception of the impression and the execution of the reaction. Thus we have a voluntary act of relatively simple character when we allow an act of cognition or discrimination to follow the perception of the impression and then let the movement depend on this second process. In this case not the immediate impression but the idea that results from the act of cognition or discrimination is the motive for the act to be performed. This motive is only one of a greater or smaller number of equally possible motives that could have come up in place of it; as a result the reaction-movement takes on the character of a voluntary act. In fact, we may observe clearly the feeling of resolution antecedent to the act and also the feelings preceding that and connected with the perception of the impression. This is still more emphatically the case, and the succession of ideational and affective processes is at the same time more complicated, when we bring in still another psychical process, as, for example, an association, to serve as the decisive motive for the execution of the movement. Finally, the voluntary process becomes one of choice when, in such experiments, the act is not merely influenced by a plurality of motives in such a way that several must follow one another before one determines the act, but when, in addition to that, one of a number of possible different acts is decided upon according to the motive presented. This takes place when preparations are made for different movements, for example, one with the right, another with the left hand, or one with each of the ten fingers, and the condition is prescribed for each movement that an impression of a particular quality shall serve as its motive, for example, the impression blue for the right hand, red for the left.

13. Muscular reactions, on the contrary, may be used follow out the *retrogradation of volitional acts* to reflex movement. In this form of reaction the preparatory expectation is directed entirely towards the external act, so that a voluntary inhibition or execution of the act in accordance with the special character of the impression, that is, a transition from simple to complex acts of will, is in this case impossible. On the other hand, it is easy by practice so to habituate one's self to the invariable connection of an impression

---

[11] The reaction-times for sensations of taste, smell, temperature, and pain are not reckoned in the figures given. They are all longer. The differences are, however, obviously to be attributed to pure physiological conditions (slow transmission of the stimulation to the nerve-endings, and in the case of pain slower central conduction), so that they are of no interest for psychology.

and a particular movement, that the process perception fades out more and more or takes place the motor impulse, and finally the movement becomes like a reflex movement. This reduction of volition to mechanical process, which in the case of sensorial reactions is never possible from the very nature of their conditions, shows itself in the shortening of the objective time to that observed for pure reflexes, and in the subjective coincidence in point of time of impression and reaction, while the characteristic feeling of resolution gradually disappears entirely.

13a. The chronometric experiments familiar in experimental psychology under the name of "reaction-experiments", are important for two reasons: first, as aids in the analysis of volitional processes, and secondly, as means for the investigation of the temporal course of psychical processes in general. This twofold importance of reaction-experiments reflects the central importance of volitions. On the one hand, the simpler processes, feelings, emotions, and their related ideas, are components of a complete volition; on the other, all possible forms of the interconnection of psychical compounds may appear as components of a volition. Volitional processes are, consequently, an appropriate transition to the interconnection between psychical compounds to be discussed in the next chapter.

For a "reaction-experiment" which is to be the basis of an analysis of a volitional process or any of its component psychical processes, we must have first of all exact and sufficiently fine (reading with exactness to 1/1000 *sec.) chronometric apparatus* (electric clock or graphic register). The apparatus must be so arranged that we can determine exactly the moment at which the stimulus acts and that at which the subject reacts. This can be accomplished by allowing the stimulus itself (sound, light, or tactual stimulus) to close an electric current that sets an electric clock reading to 1/1000 sec., in motion, and then allowing the observer, by means of a simple movement of the hand which raises a telegraph-key, to break the current again at the moment in which he apprehends the stimulus. In this way we may measure simple reactions varied in different ways (sensorial and muscular reactions, reactions with or without preceding signals), or we may bring into the process various other psychical acts (discriminations, cognitions, associations, selective processes) which may be regarded either as motives for the volition or as components of the general interconnection of psychical compounds. A simple reaction always includes, along with the volitional process, purely physiological factors (conduction of the sensory excitation to the brain and of the motor excitation to the muscle). If, now, we insert further psychical processes (discriminations, cognitions, associations, acts of choice), a modification which can be made only when sensorial reactions are employed, the duration of clearly definable psychical processes may be gained by subtracting the interval found for simple reactions from those found for the compound reactions. In this way it has been determined that the time required for the cognition and for the discrimination of relatively simple impressions (colors, letters, short words) is 0.03 - 0.05"; the time for choice between two movements (right and left hand) is 0.06", between ten movements ,the ten fingers) 0.4", etc. As already remarked, the value of these figures is not their absolute magnitude, but rather their utility as cheeks for introspection, while at the same time we may apply this introspective observation to processes subject to conditions which are prescribed with exactness by means of experimental methods and which may therefore be repeated at pleasure.

CHAPTER 3

INTERCONNECTION OF PSYCHICAL COMPOUNDS

15. CONSCIOUSNESS AND ATTENTION

1. Every psychical compound is composed of a number of psychical elements which do not usually all begin or end at exactly the same moment. As a result, the interconnection which unites the elements to a single whole always reaches beyond the individual compounds, so that different simultaneous and successive compounds are united, though indeed somewhat more loosely. We call this interconnection of psychical compounds conscious.

Consciousness, accordingly, does not mean anything that exists apart from psychical processes, nor does it refer merely to the sum of these processes without reference to how they are related to one another. It expresses the general synthesis of psychical processes, in which the single compounds are marked off as more intimate combinations. A state in which this interconnection is interrupted, as deep sleep or a faint, is called an *unconscious* state; and we speak of "disturbances of consciousness" when abnormal changes in the combination of psychical compounds arise, even though these compounds themselves show no changes whatever.

Consciousness in this sense, as a comprehensive interconnection of simultaneous and successive psychical processes, shows itself in experience first of all in the psychical life of the *individual as individual consciousness.* But we have analogous interconnection in the combination of individuals, although it is limited to certain sides of mental life, so that we may further include under the more general concept consciousness the concepts of *collective consciousness, of social consciousness,* etc. For all these broader forms, however, the foundation is the individual consciousness, and it is to this that we will first turn our attention.

Individual consciousness stands under the same external conditions as psychical phenomena in general, for which it is, indeed, merely another expression, referring more particularly to the mutual relations of the components of these phenomena to one another. As the substratum for the manifestations of an individual consciousness we have in every case an individual animal organism. In the case of men and similar higher animals the cerebral cortex, in the cells and fibers of which all the organs that stand in relation to psychical processes are represented, appears as the immediate organ of this consciousness. The complete interconnection of the cortical elements may be looked upon as the physiological correlate of the interconnection of psychical processes in consciousness, and the differentiation in the functions of different cortical regions as the physiological correlate of the great variety of single conscious processes. The differentiation of functions in the central organ is, indeed, always merely relative; every psychical compound requires the cooperation of numerous elements and many central regions. When the destruction of certain cortical regions produces definite disturbances in voluntary movements, or in sensations, or when it interferes which the formation of certain classes of ideas, it is perfectly justifiable to conclude that this region furnishes certain links in the chain of psychical elements that are indispensable for the processes in question. The assumptions often made on the basis of these phenomena, that there is in the brain a special organ for the faculties of speech and writing, or that visual, tonal, and verbal ideas are stored in special cortical cells, are not only the results of the grossest physiological misconceptions, but they are absolutely irreconcilable with the psychological analysis of these functions. Psychologically regarded, these assumptions are nothing but modern revivals of that most unfortunate form of faculty-psychology known as phrenology.

2a. The facts that have been discovered in regard to the localization of certain psycho-physical functions in the cortex, are derived partly from pathological and anatomical observations on men and partly from experiments on animals. They may be summed up as follows: 1) Certain cortical regions correspond to certain peripheral sensory and muscular regions. Thus, the cortex of the occipital lobe is connected with the retina, a part of the parietal lobe with the tactual surface, and a part of the temporal with the auditory organ. The central ganglia of special groups of muscles generally lie directly next to or between the sensory centers functionally related to them. 2) Certain complex disturbances have been demonstrated when certain cortical regions which are not directly connected with peripheral organs, but are inserted between other central regions, fail to carry out their functions. The only relation of this kind which has been proved with certainty, is that of a certain region of the temporal lobe to the functions of *speech*. The front part of this region is connected in particular with the articulation of words (its disturbance results in interference with motor coordination, so-called "ataxic aphasic"), the part further back is connected with the formation of word-ideas (its disturbance hinders sensorial coordination and produces in this way the so-called "amnesic aphasia"). It is also observed that these functions are as a rule confined entirely to the left temporal lobe and that generally apoplectic disturbances in the right lobe do not interfere with speech, while those in the left lobe do. Furthermore, in all these cases, in both simple and complex disturbances, there is usually a gradual restoration of the functions in the course of time. This is probably effected by the vicarious functioning of some, generally a neighboring cortical region in place of that which is disturbed (in disturbances of speech, perhaps it is the opposite, before untrained, side that comes into play). Localization of other complex psychical functions, such as processes of memory and association, has not yet been demonstrated with certainty. The name "psychical centers", applied to certain cortical regions by many anatomists, is for the present at least based exclusively either on the very questionable interpretation of experiments on animals, or else on the mere anatomical fact that no motor or sensory fibers running directly to these regions can be found, and that their connective fibers in general are developed relatively late. The cortex of the frontal brain is such a region. In the human brain it is noticeable for its large development. It has been observed in many cases that disturbances of this part of the brain soon result in marked inability to concentrate the attention or in other intellectual defects which are possibly reduceable to this; and from these observations the hypothesis has been made that this region is to be regarded as the seat of the function of *apperception* which will be discussed later (4), and of all those components of psychical experience in which as in the feelings, the unitary interconnection of mental life finds its expression. This hypothesis requires, however, a firmer empirical foundation than it has at present. It is to be noted that those cases where, in contrast with the first ones, mentioned, a partial injury of the frontal lobe is sustained without any noticeable disturbance of intelligence, are by no means proofs against this hypothesis. There is much evidence to show that just here, in the higher centers, local injuries may occur without any apparent results. This is probably due to the great complexity of the connections and to the various ways in which the different elements can, therefore, take the place of one another. The expression "center" in all these cases is, of course, employed in the sense that is justified by the general relation of psychical to physical functions, that is, in the sense of a parallelism between the two classes of elementary processes, the one regarded from the point of view of the natural sciences, the other from that of psychology.

3. The interconnection of psychical processes, which constitutes what we understand under the concept consciousness, is in part a simultaneous, in part a successive interconnection. The sum of all the processes present at a given moment is always a unitary whole whose parts are more or less closely united. This is the simultaneous interconnection. A present state is derived directly from that immediately preceding either through the disappearance of certain processes while others change their course and still others begin, or, when a state of unconsciousness intervenes, the new processes are brought into relation with those that were present before. These are successive interconnections. In all these cases the scope of the single combinations between preceding and following processes determines the state of consciousness. Consciousness gives place to unconsciousness when this interconnection is completely interrupted, and it is more incomplete the looser the connection of the processes of the moment with those preceding, Thus, after a period of unconsciousness the normal state of consciousness is generally only slowly recovered through a gradual reestablishment of relations with earlier experiences.

So we come to distinguish *grades* of consciousness. The lower limit, or zero grade, is unconsciousness. This condition, which consists in an absolute absence of all psychical interconnections, is essentially different from the disappearance of single psychical contents from consciousness. The latter is continually taking place in the flow of mental processes. Complex ideas and feelings and even single elements of these compounds may disappear, and new ones take their places. This continuous appearance and disappearance of elementary and composite processes in consciousness is what makes up *its successive* interconnection. Without this change, such an interconnection would, of course, be impossible. Any psychical element that has disappeared from consciousness, is to be called *unconscious* in the sense that we assume the possibility of its renewal, that in its reappearance in the actual interconnection of psychical processes. Our knowledge of an element that has become unconscious does not extend beyond this possibility of its renewal. For psychology, therefore, it has no meaning except as a *disposition* for the rise of future components of psychical processes which are connected with others before present. Assumptions as to the state of the "unconscious" or as to "unconscious processes" of any kind which are thought of as existing along with the conscious processes of experience, are entirely unproductive for psychology. There are, of course, *physical* concomitants of the psychical dispositions mentioned, of which some can be directly demonstrated, some inferred from various experiences. These physical concomitants are the effects which *practice* produces on all organs, especially those of the nervous system. As a universal result of practice we observe a *facilitation of action* which renders a repetition of the process easier. To be sure, we do not know any details in regard to the changes that are effected in the structure of the nervous elements through practice, but we can represent them to ourselves through very natural analogies with mechanical processes, such, for example, as the reduction of friction resulting from the rubbing of two surfaces against each other.

It was noted in the case of temporal ideas, that the member of a series of successive ideas which is immediately, *present* in our perception, has the most favorable position. Similarly in the simultaneous interconnection of consciousness, for example in a compound clang or in a series of new objects, certain single components are favored above the others. In both cases we designate the differences in perception as differences in *cleanness and distinctness.* Clearness, is the, relatively favorable comprehension of the object in itself distinctness the sharp discrimination from other objects, which is generally connected with clearness. The state which accompanies the clear grasp of any psychical. content and is characterized by a special feeling, we call *attention.* The process through which any such content is brought to clear comprehension we call *apperception.* In contrast with this, perception which is not accompanied by a state of attention, we designate *apprehension.* Those contents of consciousness upon which the attention is concentrated are spoken of, after the analogy of the external optical fixation point, as the *fixation-point of consciousness,* or the *inner fixation-point.* On the other hand, the whole content of consciousness at any given moment is called the *field of conscious.* When a psychical process passes into an unconscious state we speak of its *sinking below the threshold of consciousness* and when such a process arises we say it *appears above the threshold of consciousness.* These are all figurative expressions and must not be understood literally. They are useful, however, because of the brevity and clearness they permit in the description of conscious processes.

5. If we try to describe the train of psychical compounds in their interconnection with the aid of these figurative expressions, we may say that it is made up of a continual coming and going. At first some compound comes into the field of consciousness and then advances into the inner fixation-point, from which it returns to the field of consciousness before disappearing entirely. Besides this train of psychical compounds which are apperceived, there is also a coming and going of others which are merely

125

apprehended, that is, enter the field of consciousness and pass out again without reaching the inner fixation-point. Both the apperceived and the apprehended compounds may have different grades of clearness. In the case of the first class this appears in the fact that the clearness and distinctness of apperception in general is variable according to the state of consciousness. To illustrate: it can easily be shown that when one and the same impression is apperceived several times in succession if the other conditions remain the same, the successive apperceptions are usually clearer and more distinct. The, different degrees of clearness in the case of compounds that, merely apprehended, may be observed most easily when the impressions are composite. It is then found, especially when the impressions last but an instant, that even here,: where all the components are obscure from the first, that there are still different gradations. Some seem to rise more above the threshold of consciousness, some less.

6. These relations can not be determined through chance introspections, but only by systematic experimental observations. The best kinds of conscious contents to use for such observations are ideas because they can be easily produced at any time through external impressions. Now, in any temporal idea, as already remarked, those components which belong to the present moment are in the fixation-point of consciousness. Those of the preceding impressions which were present shortly before, are still in the field of consciousness, while those which were present longer before, have appeared from consciousness entirely. A spacial idea, on the, other hand, when it has only a limited extent, may be apperceived at once in its totality. If it is more composite, then its parts too must pass successively through the inner fixation-point if they are to be clearly apprehended. It follows, therefore, that composite *spacial* ideas (especially momentary visual impressions) are peculiarly well suited to furnish a measure of the amount of content that can be *apperceived* in a single act, or of the *scope of attention;* while, composite *temporal* ideas (for example, rhythmical auditory impressions, hammer-strokes) may be used for measuring the amount of all the contents that can enter into consciousness at a given moment, or the *scope of consciousness.* Experiments made in this way give, under different conditions, a scope of from 6 to 12 simple impressions for attention and of 16 to 40 such impressions for consciousness. The smaller figures are for those impressions which do not unite at all to ideational combinations, or at most very incompletely, the larger for those in which the elements combine as far as possible to composite ideas.

6 a. The most accurate way of determining the *scope of attention* is to use spacial impressions of sight, for in such cases it is very easy, by means of an electric spark, or the fall of a screen made with an opening in the center, to expose the objects for an distant and in such a way that they all lie in the region of clearest vision. This gives us physiological conditions that do not prevent the apperception of a greater number of impressions than it is possible to apperceive because of the limited scope of attention. In these experiments there must be a point for fixation in the middle of the surface on which the impressions lie, before the momentary illumination. Immediately after the experiment, if it is properly arranged, the observer knows that the number of objects which were clearly seen in a physiological sense, is greater than the number included within the scope of attention. When, for example, a momentary impression is made up

of letters, it is possible, by calling up a memory-image of the impression, to read afterwards some of the letters that were only indistinctly apprehended at the moment of illumination. This memory-image, however, is clearly distinguished in time from the impression itself, so that the determination of the scope of attention is not disturbed by it. Careful introspection easily succeeds in fixating the state of consciousness at the moment the impression arrives, and in distinguishing this from the subsequent acts of memory, which are always separated from it by a noticeable interval. Experiments made in this way show that the scope of attention is by no means a constant magnitude, but that, even when the concentration of the attention is approximately at its maximum, its scope depends in part on the simplicity or complexity of the impressions, in part on their familiarity. The simplest spacial impressions are arbitrarily distributed points. Of these a maximum of six can be apperceived at one time. When the impressions are somewhat more complex but of a familiar character, such as simple lines, figures, and letters, three or four of them are generally apperceived simultaneously, or, under favorable conditions, even five. The figures just given hold for vision; for touch the same limits seem to hold only in the case of points. Six such simple impressions can, under favorable conditions, be apperceived in the same instant. When the impressions are familiar but complex, even for vision, the number of ideas decreases, while that of the single elements increases very markedly. Thus, we can apperceive two or even three familiar monosyllabic words, which contain in all ten or twelve single letters. Under any circumstances, then, the assertion often made, that the attention can be concentrated on only one idea at a time, is false.

Then, too, these observations overthrow the assumption sometimes accepted, that the attention can sweep continuously and with great rapidity over a great number of single ideas. In the experiment described, if the attempt is made to fill up from memory the image which is clearly perceived an instant after the impression, a very noticeable interval is required to bring into clear consciousness an impression that was not apperceived at first; and in the process the first image always disappears from attention. The successive movement of attention over a number of objects is, accordingly, a discontinuous process, made up of a number of separate acts of apperception following one another. This discontinuity is due to the fact that every single apperception is made up of a period of increasing followed of by a period decreasing strain. The period of maximal tension between the two, may vary considerably in its duration. In the case of momentary, and rapidly changing impressions, it is very brief; when, on the other hand, we concentrate on particular objects, it is longer. But, even when the attention is thus concentrated on objects of a constant character, a periodic interruption, due to the alternating relaxation and renewed concentration, always appears. This may be easily observed, even in the ordinary action of attention. But here, too, we gain more detailed information through experiments. If we allow a weak, continuous impression to act on a sense-organ and remove so far as possible all other stimuli, it will be observed when the attention is concentrated upon it that at certain, generally irregular, intervals the impression becomes for a short time indistinct, or even appears. to fade out entirely, only to appear again the next moment. This wavering begins, when the impressions are very weak, after 3-6"; when they are somewhat stronger, after 18-24". These variations are readily distinguished from changes in the intensity of the impression itself, as may be easily demonstrated when, in the course of the experiment, the stimulus is purposely weakened or interrupted. There are essentially two characteristics that distinguish the subjective variations from those due to the changes in the stimulus. First, so long as the impression merely passes back and forth from the obscure field of consciousness to the inner fixation-point, there is always an idea of its continuance, just as there was in the experiments with momentary impressions an indefinite and obscure idea of the components which were not apperceived. Secondly, the oscillations of attention are attended by characteristic feelings and sensations which are entirely absent when the changes are objective. The characteristic feelings are those of expectation and activity, which regularly increase with the concentration of attention and decrease with its relaxation. These will be discussed more fully later. The sensations come from the sense-organ affected, or at least emanate indirectly from it. They consist in sensations of tension in the tympanum, or in those of accommodation and convergence, etc. These two series of characteristics distinguish the concepts of the clearness and distinctness of psychical contents from that of the intensity of their sensational elements. A strong impression may be obscure and a weak one clear. The only causal relation between these two different concepts is to be found in the fact that in general the stronger impressions force themselves more upon the apperception. Whether or not they are really more clearly apperceived, depends on the other conditions present at the moment. The same is true of the advantages which those parts of a visual impression have that fall within the region of clearest vision. As a rule, the fixated objects are also the ones apperceived. But, in the experiments with momentary impressions described above, it can be shown that this interconnection may be broken up. This happens when we voluntarily concentrate our attention on a point in the eccentric regions of the field of vision. The object which is *obscurely seen* then becomes the one which is *clearly ideated.*

6b. In the same way that momentary spacial impressions are used to determine the scope of attention, we may use those which succeed one another in time, as a measure for the *scope of consciousness.* In this case we start with the assumption that a series of impressions can be united in a single unitary idea only when they are all together in consciousness, at least for one moment. If we listen to a series of hammer-strokes, it is obvious that while the present sound is apperceived, those immediately preceding it are still in the field of consciousness. Their clearness diminishes, however, just in proportion to their distance in time from the apperceived impression, and those lying beyond a certain limit disappear from consciousness entirely. If we can determine this limit, we shall have a direct measure for the scope of consciousness under the special conditions given in the experiment. As a means for the determination of this limit we may use the ability to compare temporal ideas that follow one another immediately. So long as such an idea is present in consciousness as a single unitary whole, we can compare a succeeding idea with it and decide whether the two are alike or not. On the other hand, such a comparison is absolutely impossible when the preceding temporal series is not a unitary whole for consciousness, that is, when a part of its constituents have passed into unconsciousness before the end is reached. If, then, we present two series of strokes, such as can be produced, for example, by a metronome, one immediately after the other, marking of each series by a signal at its beginning, as, for example, with a bell stroke, we can judge directly from the impression, so long as they can be grasped as single units in consciousness, whether the, two series are alike or not. Of course, in such experiments counting of the strokes must be strictly avoided. judgments it may be noticed that the impression produced by the affective elements of the temporal before. Every stroke in the second series is preceded by a feeling of expectation corresponding to the analogous stroke of the first series, so that every stroke too many or too few produces a feeling of disappointment attending the disturbance of the expectation. It follows that it is not necessary for the two successive series to be present in consciousness at the same time in order that they may be compared; but what is required is the union of all the impressions of one series together in a single unitary idea. The relatively fixed boundary of the scope of consciousness is clearly shown in the fact that the likeness of two temporal ideas is always recognized with certainty so long as they do not pass the bound that holds for the conditions under which they are given, while the judgment becomes absolutely uncertain when this limit is once crossed. The extent of the scope of consciousness as found in measurements made when the conditions of attention remain the same, depends partly on the rate of the successive impressions and partly on their more or less complete rhythmical combination. When the rate of succession is slower than about 4", it becomes impossible to combine successive impressions to a temporal idea; by the time a new impression arrives, the preceding one has already disappeared from consciousness. When the rate passes the upper limit of about 0.18", the formation of distinctly defined temporal ideas is impossible because the attention can not follow the impressions any longer. The most favorable rate is a succession of strokes every 0.2~0.3". With this rate, and with the simplest rhythm, which generally arises of itself when the perception is uninfluenced by any special objective conditions, the 2/8-time 8 double or 16 single impressions can be just grasped together. The best measure for the apprehension of the greatest possible number of single impressions is the 4/4 measure with the strong accent on the first stroke and the medium accent on the fifth. In this case a maximum of five feet or forty single impressions can be grasped at once. If these figures are compared with those obtained when the scope of attention was measured, putting simple and compound temporal impressions equal to the corresponding spacial impressions, we find that the scope of consciousness is about four times as great as that of attention.

7. Besides the properties of clearness and distinctness, which belong to conscious contents in themselves or in their, mutual relations to one another, there are regularly others which are immediately recognized as *accompanying* processes. These are partly affective processes that are characteristic for particular forms of apprehension and apperception, partly, sensations of a somewhat variable character. Especially the ways in which psychical contents *enter* the field and fixation-point of consciousness vary according to the different conditions under which this entrance may take place. When any psychical process rises above the threshold of consciousness, the affective elements, as soon as they are strong enough, are what first become noticeable. They begin to force themselves energetically into the fixation-point of consciousness before anything is perceived of the ideational elements. This is the case whether the impressions are new or revivals of earlier processes. This is what causes the peculiar states of mind which we are not exactly able to account for, some-times of a pleasurable or unpleasurable

character, sometimes predominantly states of strained expectation. In this last case the sudden entrance of the ideational elements belonging to the feelings, into the scope of the attention, is accompanied by feelings of relief or satisfaction. When we are trying to recall something that has been forgotten, the same affective state may arise. Often there is vividly present in such a case, besides the regular feeling of strain, the special affective tone of the forgotten idea, although the idea itself still remains in the background of consciousness. In a similar manner, as we shall see later (§ 16), the clear apperception of ideas in acts of cognition and recognition is always preceded by special feelings. Similar affective states may be produced experimentally by the momentary illumination of a field of vision in which there are impressions of the strongest possible affective tone in the region of indirect vision. All these experiences seem to show that every content of consciousness has some influence on attention. It shows this regularly in its own affective coloring, partly in the feelings regularly connected with acts of attention. The whole effect of these obscure contents of consciousness on the attention fuses, according to the general law of the synthesis of affective components, with the feelings attending the apperceived contents to form a single total feeling.

8. When psychical content enters the *fixation point of* consciousness, new and peculiar affective processes are added to those that have been described. These new feelings may be of a variety of kinds, according to the different conditions attending this entrance into the fixation-point. The conditions are of *two* classes, and are interconnected for the most part with the above described preparatory affective influences of the content not yet apperceived.

First, the new content forces itself on the attention suddenly and without preparatory affective influences; this we call *passage apperception.* While the content of consciousness is becoming clearer both in its ideational and affective elements, there is first of all a concomitant feeling of *passive receptivity,* which is a depressing feeling, and generally stronger the more intense the psychical processes, and the more rapid its rise. This feeling soon sinks and then gives place to an antagonistic, exciting feeling of *activity.* There are connected with both these feelings characteristic sensations in the muscles of the sense-organ from which the ideational components of the process proceed. The feeling of receptivity is generally accompanied by a transient sensation Of relaxation, that of activity by a succeeding sensation of strain.

Secondly, the new content is preceded by the preparatory affective influences mentioned above (7), and as a result the attention is concentrated upon it even before it arrives; this we call *active apperception.* In such a case the apperception of the content is preceded by a feeling of *expectation,* sometimes of longer, sometimes of shorter duration. This feeling is generally one of strain and may at the same time be one of excitement; it may also have pleasurable or unpleasurable factors, according to its ideational elements. This feeling of expectation is usually accompanied by fairly intense sensations of tension in the muscles of the sense-organ affected. At the moment in which the content arises in clear consciousness, this feeling gives place to a feeling of fulfillment which is generally very short and always has the character of a feeling of relief. Under circumstances it may also be depressing or exciting, pleasurable or unpleasurable. After this feeling of fulfillment we have at once that of activity - the same that appeared at the close of passive apperception, and is here, too, united with an increase in feelings of strain.

8a. The experimental observation of the different forms of apperception can be carried out best with the aid of the reaction-experiments described in § 14, 11 sq. Passive apperception may be studied by the use of unexpected, and active by the use of expected impressions. At the same time it will be observed that between these typical differences there are intermediate stages. Either the passive form will approach the active because of the weakness of the first stage, or the active will approach the passive form because in the sudden relaxation of the expectation the contrast between the expectation and the relief and depression which come in the succeeding feeling of fulfillment, is more marked than usual. In reality we have everywhere continuously interconnected processes which are opposite character only in extreme cases.

9. If the affective side of these processes of attention axe more closely examined, it is obvious that they are exactly the same as the affective content of all *volitional processes.*

At the same time it is clear that in its essential character passive apperception corresponds to a simple impulsive act, while the active form corresponds to a complex voluntary act. In the first case we may evidently regard the psychical content that forces itself upon attention without preparation, as the single motive which, without any conflict with other motives, gives rise to the act of apperception. The act is here too connected with the feeling of activity characteristic of all volitional acts. In the case of active apperception, on the other hand, other psychical contents with their affective elements tend to force themselves upon the attention during the preparatory affective stages, so that the act of apperception when it finally is performed is often recognized as a voluntary process or even as a selective process when the conflict between different contents comes clearly into consciousness. The existence of such selective acts under the circumstances mentioned was recognized even in older psychology where "voluntary attention" was spoken of. But here too, as in the case of external volitional acts, will stood alone; there was no explanation of it by its antecedents, for the central point in the development, namely, the fact that so-called involuntary attention is only a simpler form of internal volition, was entirely overlooked. Then, too, in full accord with the methods of the old faculty-theory "attention" and "will" were regarded as different, sometimes as related, sometimes as mutually excluding psychical forces, while the truth evidently is that these two concepts refer to the same class of psychical processes. The only difference is that processes of apperception and attention are those which occur only as so-called internal acts, that is, have no external effects except indirectly when they lead to other processes.

10. Connected with these internal volitional acts, which we call processes of attention, there takes place a formation of certain concepts of the highest importance for all psychical, development. This is the formation of the concept *subject* and the correlate presupposition of *objects* as independent realities standing over against the subject. This

can be carried out in its logical form only with the aid of scientific reflection, still it has its substratum in the processes of attention.

Even in immediate experience there is a division between components of this experience. On the one hand are those which are arranged in space with relation to the point of orientation mentioned above, and are called either *objects,* that is, something outside the perceiving subject, or, when we attend to the mode of their rise in consciousness, ideas, that is something which the subject perceives. On the other hand, there are those contents of experience which do not belong to this spacial order, though they are continually brought into relation with it through their quality and intensity. These latter contents, as we saw in § 12-14, are intimately interconnected. *Feelings* are parts of *emotions* and emotions are to be considered as components of *volitional processes.* The process, may end before it is fully completed, as often when a feeling gives rise to no noticeable emotion, or when an emotion fades out without really causing the volitional act for which prepared the way. All these affective processes may, accordingly, be subsumed under the general concept *volitional process.* This is the complete process of which the two others are merely components of simpler or more complex character. From this point of view we can easily understand how it is that even simple feelings contain, in the extremes be they vary, a volitional direction; and express, in the same way the amount of volitional energy present at a given moment; and finally, correspond to certain particular phases of the volitional process itself. The *direction of volition* is obviously indicated by the pleasurable or unpleasurable directions of feelings, which correspond directly to some sort of effort to reach something or to avoid it. The *energy of volition* finds its expression in the arousing and subduing directions of feelings, while the opposite *phases* of a volitional process are related to the directions of strain and relaxation.

11. Thus, *volition* proves to be the fundamental fact from which all those processes arise which are made up of feelings. Then, too, in the process of *apperception,* which is found through psychological analysis to have all the characteristics of a volitional act, we have a direct relation between this fundamental fact and the ideational contents of experience which arise from the spacial arrangement of sensations. Now, volitional processes are apprehended as unitary processes and as being uniform in character in the midst of all the variations in their components. As a result there arises an immediate feeling of this unitary interconnection, which is most intimately connected with the feeling of activity that accompanies all volition, and then is carried over to all conscious contents because of their relation to will, as mentioned above. This feeling of the interconnection of all single psychical experiences is called the "ego". It is a *feeling,* not an idea as it is often called. Like all feelings, however, it is connected with certain sensations and ideas. The ideational components most closely related to the ego are the common sensations and the idea of one's own body.

That part of the affective and ideational contents which separates off from the totality of consciousness and fuses closely with the feeling of the ego, is called *self-consciousness.* It is no more a reality, apart from the processes of which it is made up, than is consciousness in general, but merely Points out the interconnection of these processes,

132

which furthermore, especially in their ideational components, can never be sharply distinguished from the rest of consciousness. This shows itself most of all in the fact that the idea of one's own body sometimes fuses with the feeling of the ego, sometimes is distinct from it as the idea of an object, and that in general self-consciousness in its development always tends to reduce itself to its affective basis.

12.This separation of self-consciousness from the other contents of consciousness also gives rise to the discrimination *of subject and objects.* This discrimination was prepared for, to be sure, by the characteristic differences among the original contents of consciousness, but is fully carried out only as a consequence of this separation. The concept *subject* has accordingly as a result of its psychological development three different meanings of different scope, each of which may at different times be the one employed. In its narrowest sense the subject is the interconnection of volitional processes which finds expression in the feeling of the ego. In the next wider sense it includes the real content of these volitional processes together with the feelings and emotions that prepare their way. Finally, in its widest significance it embraces the constant ideational substratum of these subjective processes, that is, the body of the individual as the seat of the common sensations. In the line of development the widest significance is the oldest, and in actual psychical experience the narrowest is continually giving way to a return of one of the others because it can be fully attained only through conceptual abstraction. This highest form is, then, in reality merely a kind of limits towards which the self-consciousness may approach more or less closely.

12a. This discrimination of subject and objects, or the *ego* and the outer *world* as it is commonly expressed by reducing first concept to its original affective substratum and the second together in a general concept - this discrimination of all the considerations responsible for the dualism which first gained currency in the popular view of things and was then carried over into the philosophical systems. It is on this ground that psychology comes to be set over against the other sciences, in particular the natural sciences, as a science of the subject (§ 1, 3a.) This view could be right only under the conditions that the discrimination of the ego from the outer world were a fact preceding all experience and that the concepts subject and objects could be unequivocally distinguished once for all. But neither of these conditions is fulfilled. Self-consciousness depends on a whole series of psychical processes of which it is the product, not the producer. Subject and object are, therefore, neither originally nor in later development absolutely different contents of experience, but they are concepts which are due to the reflection resulting from the interrelations of the various components of the absolutely unitary content of our immediate experience.

13. The interconnection of psychical processes which makes up consciousness, necessarily has its deepest spring in the *processes of combination* which are continually taking place between the elements of the single contents of experience. Such processes are operative in the formation of single psychical compounds and they are what give rise to the simultaneous unity of the state of consciousness present at a given moment and also to the continuity of successive states. These processes of combination are of the most various kinds; each one has its individual coloring, which is never exactly

reproduced in any second case. Still, the most general differences are those exhibited by the attention in the passive reception of impressions and the active apperception of the same. As short names for these differences we use the term *association* to indicate a process of combination in a passive state of attention, and *apperceptive combination* to indicate a combination in which the attention is active.

16. ASSOCIATIONS

1. The concept association has undergone, in the development of psychology, a necessary and very radical change in meaning. To be sure, this change has not been accepted everywhere, and the original meaning is still retained, especially by those psychologists who support, even today, the fundamental positions on which the association-psychology grew up. This psychology is predominantly intellectualistic, pays attention to nothing but the *ideational contents* of consciousness and, according limits the concept of association to the combinations of ideas. Hartley and Hume, the two founders of association-psychology, spoke of "association of ideas" in this limited sense.[12] Ideas were regarded as objects, or at least as processes that could be repeated in consciousness with exactly same character as that in which they were present at first. This led to the view that association was a principle for the explanation of the so-called "reproduction" of ideas. Furthermore, it was not considered necessary to account for the rise of composite ideas with the aid of psychological analysis, since it was assumed that the physical union of impressions in sense-perception was sufficient to explain the psychological composition and so the concept of association was limited to those forms of so-called reproduction in which the associated ideas succeed one another in time. For the discrimination of the chief forms of successive associations Aristotle's logical scheme for the memory-processes was accepted, and in accordance with the principle classification by opposites the following forms were discriminated: association by similarity and contrast, and association by simultaneity and succession. These class-concepts gained by a logical dichotomic process were dignified with the name of "law of associations". Modern psychology has generally sought to reduce the number of these laws. Contrast is as a special form of similarity, for only those concepts are associated which belong to the same class; and associations by simultaneity and succession included under contiguity. *Contiguity* is then regarded as *outer* association and contrasted with *inner* association by *similarity*. Some psychologists believe it possible to reduce two forms to a single, still more fundamental, "law of association" by making association by contiguity a special form of similarity what is still more common, by explaining similarity as a result of association by contiguity. In both cases association is generally brought under the more general idea of practice or habituation.

---

[12] The author [Wundt] remarks that the English word idea as here used corresponds to the German *Vorstellung*. Tr. [Judd]

2. The whole foundation for this kind of theorizing is destroyed by *two* facts which force themselves irresistibly upon us as soon as we begin to study the matter experimentally. The *first* of these facts is the general result of the psychological analysis of sense-perceptions, that composite ideas, which association-psychology regards as irreducible psychical units, are in fact the results of synthetic processes which are obviously in close interconnection with the processes commonly called associations. The *second* fact comes from the experimental investigation of memory-processes. It is found that the *reproduction* of ideas in the strict sense of a renewal in its unchanged form of an earlier idea, takes place at all, but that what really does happen of memory is the rise of a new idea in consciousness, always differing from the earlier idea to which it is referred, and deriving its elements as a rule from various preceding ideas.

It follows from the first fact that there are elementary processes of association between the components of ideas preceding the associations of composite ideas with one another which the name is generally limited. The second fact proves that ordinary associations can be nothing but complex products of such elementary associations. These can show the utter unjustifiableness of excluding the elementary processes whose products are simultaneous ideas rather than successive, from the concept association. Then, too, there no reason for limiting the concept to ideational processes. The existence of composite feelings, emotions, etc., shows, on the contrary, that affective elements also enter into regular combinations, which may in turn unite with associations of sensational elements to form complex products, as we saw in the rise of *temporal ideas.* The intimate relation between the various orders of combining processes and the necessity of elementary associations as antecedents to all complex combinations, furnishes further support for the observation made on the general mode of the occurrence of conscious processes, that it is never possible to draw a sharp boundary line between the combinations of the elements that compose psychical compounds, and the interconnection of the various psychical compounds, in consciousness.

3. It follows that the concept of association can gain a fixed, and in any particular case unequivocal, significance, only association is regarded as an *elementary process* which never shows itself in the actual psychical processes except in and or less complex form, so that the only way to find out character of elementary association is to subject its complex products to a psychological analysis. The ordinarily so-called associations (the successive associations) are only one, and loosest at that, of all the forms of combination. In contrast with these we have the closer combinations from which the different kinds of psychical compounds arise and to which we apply the general name *fusions,* because of the closeness of the union. The elementary processes from which the compounds, the intensive, spacial, and temporal ideas, composite feelings, the emotions, and the volitional processes arise, are, accordingly, to be considered as associative processes. For the purpose of practical discrimination, however, it will be well to limit the word "association" to those combining processes which take place between elements of *different* compounds. This narrower meaning which we give the term in contrast with fusion, is in one respect an approach to the meaning that it had in older psychology for it refers exclusively to the interconnection of compounds in consciousness. It differs from the older concept, however, in two important

135

characteristics. First it is here regarded as an *elementary process,* or, when we are dealing with complex phenomena, as a product of such elementary processes. Secondly, we recognize, just as in the fusions, simultaneous associations as well as successive. In fact, the former are to be looked upon as the earlier.

A. SIMULTANEOUS ASSOCIATIONS

4. Simultaneous associations made up of elements from psychical compounds may be divided into *two* classes: into *assimilations,* or associations between the elements of *like* compounds, and *complications,* or associations elements of *unlike* compounds. Both may take place, in accordance with our limitation of the concept association, between those compounds only which are themselves simultaneous combinations, that is, between intensive and spacial ideas between composite feelings.

a. *Assimilations*

5. Assimilations are a form of association that is continually met with, especially in the case of intensive spacial ideas. It is an essential supplement to the process of formation of ideas by fusion. In the case of composite feelings this form of combination never seems to appear except where we have at the same time an assimilation of ideational elements. It is most clearly demonstrable with certain single components of the product of an assimilation given through external sense-impressions, while others believe to earlier ideas. In such a case the assimilation may be demonstrated by the fact that certain components of ideas which are wanting in the objective impression or are there represented by components other than those actually present in the idea itself, can be shown to arise from ideas. Experience shows that of these reproduced components are those are most favored which are very frequently present. Still, certain single elements of the impression are usually of more importance in determining the association than others are, so that when these dominating elements are altered, as may be the case especially with assimilation of the visual sense, the product of the assimilation undergoes a corresponding change.

6. Among intensive compounds it is especially the *auditory ideas* which are very often the results of assimilation. They also furnish the most striking examples for the principle of frequency mentioned above. Of all the auditory ideas the most familiar are the readily available *ideas of words,* for these are attended to more than other sound-impressions. As a result the hearing of words is continually accompanied by assimilations; the sound-impression is incomplete, but is entirely filled out by earlier impressions, so that we do not notice the incompleteness. So it comes that not the correct hearing of words, but the *misunderstanding* of them, that is, the erroneous filling out of incomplete impressions through incorrect assimilations, is what generally

leads us to notice the process. We may find an expression of the same fact in the ease with which any sound whatever, as, for example, the cry of an animal, the noise of water, wind, machinery, etc., can be to sound like words almost at will.

7. In the case of *intensive feelings* we note the presence of assimilations in the fact that impressions which are accompanied by sense-feelings and elementary aesthetic feelings, very exercise a second direct affective influence for which account only when we recall certain ideas of which we are reminded by the impressions. In such cases the association is usually at first only a form of affective association and only so long as this is true is the assimilation simultaneous. The ideational association which explains thesis, on the contrary, a later process belonging to the forms of successive association. For this reason it is hardly possible, when we have clang-impressions or color-impressions accompanied by particular feelings, or when we have simple spacial ideas, to decide what the immediate affective influence impression of itself is and what is that of the association. As a rule, in such cases the affective process is to be looked upon as the resultant of an immediate and an associative factor which unite to form a single, unitary total feeling in accordance with the general laws of affective fusion.

8. Association in the case of *spacial* ideas is of the most comprehensive character. It is not very noticeable in the sphere of *touch* when vision is present, on account of the importance of tactual ideas in general and especially for memory. For the blind, on the other hand, it is the means for the rapid orientation in space which is necessary, for example, in the rapid reading of the blind-alphabet. The effects of assimilation are most strikingly evident when several tactual surfaces are concerned, because in such cases its presence is easily betrayed by the illusions which arise in consequence of some disturbance in the usual relation of the sensations. Thus, for example, when we touch a small, ball with the index and middle fingers crossed, we have the idea of *two* balls. The explanation is obvious. In the ordinary position of the fingers the external impression here given actually corresponds to two balls, and the many perceptions of this kind that have been received before exercise an assimilative action on the new impression.

9. In *visual* sense-perceptions assimilative processes play a very large part. Here they aid in the formation of ideas of magnitude, distance, and three-dimensional character of visual objects. In this last respect they are essential supplements of immediate binocular motives for projection into depth. Thus, the correlation that exists between the ideas of the distance and magnitude of objects, as, for example, the apparent differing the size of the sun or moon on the horizon and at the zenith, is to be explained as an effect of assimilation. The perspective of drawing and painting also depends on these influences. A picture drawn or painted on a plane surface can appear three-dimensional only on condition that the impression arouses earlier three-dimensional ideas which are always with the new impression. The influence of these assimilation most evident in the case of unshaded drawings that can be either in relief or in intaglio. Observation shows that these differences in appearance are by no means accidental or depend upon the so-called "power of imagination", but that there are always elements in the immediate impression which determine completely the assimilative process. The elements that thus operative are, above all, the sensations arising from the position and movements of the

eye. Thus, for example, a design which can be interpreted as either a solid or a hollow prism, is seen alternately in relief and in intaglio according as we fixate in the two cases the parts of the which correspond ordinarily to a solid or to a hollow object. A solid angle represented by three lines in the same appears in relief when the fixation-point is moved along of the lines, starting from the apex, it appears in intaglio when the movement is in the opposite direction, from the of the line towards the apex. In these and all like cases assimilation is determined by the rule that in its movement the fixation-lines of objects the eye always passes from nearer to more distant points.

In other cases the geometric optical illusions, which are due to the laws of ocular movements, produce certain ideas of distance, and these not infrequently eliminate the contradictions brought about in the by the illusions. Thus, to illustrate, an interrupted straight line appears longer than an equal uninterrupted line; as a result we tend to project the first to a greater depth than the latter. Here both lines cover just the same distances on the retina in spite of the fact that their length is perceived as different, because of the different motor energy connected with their estimation. An elimination of this contradiction is effected by means of the different ideas of distance, for when one of two lines whose retinal images are appears longer than the other, it must, under the ordinary conditions of vision, belong to a more distant object. Again, one straight line is intersected at an acute angle by another, the result is an overestimation of the acute angle, sometimes gives rise, when the line is long, to an apparent bending near the point of intersection. Here contradiction between the course of the line and the increase in the size of the angle of intersection, is often eliminated by the apparent extension of the line in the third dimension. In all these cases the perspective can be explained only as the assimilative effect of earlier ideas of corresponding character.

10. In none of the assimilations discussed is it possible to show that any former idea has acted as a whole the new impression. Generally this is impossible because we must attribute the assimilative influence to a large number of ideas, differing in many respects from one another. Thus, for example, a straight line which intersects a vertical at an acute angle, corresponds to innumerable cases in which an inclination of the line with its accompanying increase of angle appeared as a component of a three-dimensional idea. But all these cases may have been very different in regard to the size of the angle, the length of the lines, and other attending circumstances. We must, accordingly, think of assimilative process as a process in which not a single definite idea or even a definite combination of elements from ideas, but as a rule a great number of such combinations are operative. These need agree only approximately with the new impression in order to affect consciousness.

We may gain some notion of the way in which this effect is produced from the important part that certain elements connected with the impression play in the process, for example, the sensations of ocular position and movements in visual ideas. Obviously it is these immediate sensational elements that serve to pick out from the mass of ideational elements which react upon the impression, certain particular ones that correspond to themselves, then bring these selected factors into a form agreeing with that of the rest of the components of the immediate impression. At the same time it appears that not merely the elements of our memory-images are relatively indefinite and therefore variable, but that even the perception of indefinite impression may vary under special conditions fairly wide limits. In this way the assimilative process starts primarily from elements of the immediate impression, chiefly from particular ones which are of preeminent importance for the formation of the idea, as, for example, the sensations of ocular position and movement in visual ideas. These elements call up certain particular memory-elements corresponding to themselves. These memories then exercise an effect on the immediate impression, and the impression in turn reacts in the same way on the reproduced moments. These separate acts are, like the whole process, not successive, but, at least for our consciousness, simultaneous. For this reason the product of the assimilation is apperceived immediate, unitary idea. The two distinguishing characteristics of assimilation are, accordingly, 1) that it is made up of a series of *elementary* processes of combination, that is, processes that have to do with the components of ideas, not with the whole ideas themselves, and 2) that the united components modify one another through *reciprocal assimilations.*

11. On this basis we can explain without difficulty the differences between complex assimilative processes, by the very different parts that the various factors necessary to such a process play in the various concrete cases. In ordinary sense-perceptions the direct elements are so predominant that the reproduced elements are as a rule entirely overlooked, although in reality they are never absent and are often very important for the perception of the objects. These reproduced elements are much more noticeable when the assimilative effect of the direct elements is hindered through external or internal influences, such as indistinctness of the impression or affective and emotional excitement. In all cases where the difference between the impression the idea becomes, in this way, so great that it is apparent once on closer examination, we call the product of assimilation an *illusion.*

The universality of assimilation makes it certain that such processes occur also between reproduced elements, in such a way that any memory-idea which arises in our mind is immediately modified by its interaction with other memory-elements. Still, in such a case we have, of course, no means of demonstration. All that can be established as probable is that even in the case of so-called "pure memory-processes" direct elements in the form of sensations and sense-fee aroused by peripheral stimuli, are never entirely absent reproduced visual images, for example, such elements present in the form of sensations of ocular position and movement.

b. *Complications*

12. *Complications,* or the combinations between psychical compounds, are no less regular components of consciousness than are assimilations. Just as there is hardly intensive or extensive idea or composite feeling which imodified in some way through the processes of reciprocal assimilation with memory-elements, so almost every one of

these compounds is at the same time connected with other, dissimilar compounds, with which it has some constant relations. In all cases, however, complications are different from assimilations in the fact that the unlikeness of the comp makes the connection looser, however regular it may be, so that when one component is direct and the other reproduced, the latter can be readily distinguished at once. Still, is another reason which makes the product of a complication unitary in spite of the easily recognized difference between its components. This cause is the *predominance of one* of the compounds, which pushes the other components into the obscurer field of consciousness.

If the complication unites a direct impression with memory-elements of disparate character, the direct impression assimilations is regularly the predominant component while the reproduced elements sometimes have a notice-able influence only through their affective tone. Thus, when we speak, the auditory word-ideas are the predominant components, and in addition we have as obscure direct motor sensations and reproductions of images of the words. In reading, on the other hand, the visual images come to the front while the rest become weaker. In general it may be said that the existence implication is frequently noticeable only through the coloring of the total feeling that accompanies the lent idea. This is due to the ability of obscure ideas to have a relatively intense effect on the attention through active tones. Thus, for example, the impression of a rough surface, a dagger-point, arises from a complication of *visual* and tactuals.

B. SUCCESSIVE ASSOCIATIONS

13. Successive association is by no means a process that differs essentially from the two forms of simultaneous association, assimilation and complication. It is, on the contrary, due to the same general causes as these, and differs only in the secondary characteristic that the process of combination, which in the former cases consisted, so far as immediate introspection was concerned, of a single instantaneous act, is here protracted and may therefore be readily divided *two* acts. The first of these acts corresponds to the appearance the *reproducing* elements, the second to the appearance the *reproduced* elements. Here too, the first act is often introduced by an external sense-impression, which a rule immediately united with an assimilation. Other reproductive elements which might enter into an assimilation or complication are held back through some inhibitory influence or other - as, for example, through other assimilations that force themselves earlier on apperception - and do not begin to exercise an influence until later. In this way have a second act of apperception clearly distinct from first, and differing from it in sensational content the more essentially the more numerous the new elements are added through the retarded assimilation and complication and the more these new elements tend to displace the earlier because of their different character.

14. In the great majority of cases the association formed is limited to *two* successive ideational or affective processes connected, in the manner described, through assimilations or complications. New sense-impressions or apperceptive combinations (§ 17) may then connect themselves with the second member of the association. Less frequently happens that the same processes which led to the first division of an assimilation or complication into a successive process, may be repeated with the second or even with the third member, so that in this way we have a whole *associational series.* Still, this takes place generally only under exceptional conditions, especially when the normal course of apperception has been disturbed, as, for example, in the so-called "flight of ideas" of the insane. In normal cases such as serial associations, that is, associations with more than two members, hardly ever appear.

14a. Such serial associations may be produced most easily under the artificial conditions of experimentation, when the effort is purposely made to suppress new sensible impressions and apperceptive combinations. But the process resulting in such cases differs from that described above in that the successive members of the series do not connect, each with its immediate predecessor, but all go back to the first, until a new sense-impression or an idea with an especially strong affective tone furnishes a new starting point for the succeeding associations. The associations-in the "flight of ideas" of the insane generally show the type of returning to certain predominant centers.

a. *Sensible Recognition and Cognition.*

The way in which the ordinary form of association, made up of two partial processes, may be most clearly observed, is in the simultaneous assimilations and complications of sensible *recognition* and *cognition.* The qualification "sensible" is when referring to these associative processes, to indicate, on the one hand, that the first member of the pro-always a sense-impression, and, on the other, to distinguish these from the logical processes of cognition.

The psychologically simplest case of recognition is that an object has been perceived - for example, seen - only *once* and is recognized as the same when met a second. If this second perception follows very soon after the first, or if the first was especially emphatic and exciting, the association usually takes place immediately as a simultaneous assimilation. This process differs from other assimilation, which take place in connection with every sense-perception, only in the characteristic accompanying feeling, *of familiarity.* Such a feeling is never present except when there is some degree of "consciousness" that the impression has already been received before. It is, therefore, evidently one of those feelings which comes from the ideas obscurely present in consciousness. The psychological difference between this and an ordinary simultaneous assimilation must be looked for in the fact that at the moment when, in the apperception of the impression, the assimilation takes place, there arise in the obscure regions of consciousness some components of the original idea which do not enter into the assimilation. Their relation to the elements of the idea that is apperceived finds expression in the feeling of familiarity. The unassimilated components may be elements of the earlier impression that were so different from certain elements of the new that they could not be assimilated, or, and this is especially often the case, they may be complications that were clear before, but now remain unobserved. This influence of complication explains how it is that the name of a visual object, for example the proper names of persons, and often other auditory qualities, such as the tone of voice, are very great helps in the recognition. To serve as such helps, however, they need not necessarily be clear ideas in consciousness. When we, have heard a man's name, the recognition of the man the next time we meet him may be aided by the name without our calling it clearly to mind.

15a. This influence of complications may be demonstrated experimentally. If we take a number of disks that are alike in all other respects, but differ in color from white through various shades of grey to black, and present them to view once, so long as only *five,* shades are used (white, black, and three shades of grey) each disk can be easily recognized again. But when more shades are used, this is no longer possible. It is very natural to surmise that this fact is related to the existence of five familiar names, white, light grey, grey, dark grey and black. This view is confirmed by the fact that by purposely using a larger number of names more shades (even as many as nine) axe recognized. In such experiments the complication may be clearly observed, but it is not necessarily so, especially for the five ordinary shades. As a rule the name is here thought of after the act of recognition proper is passed.

16. The observations discussed also show what the conditions are under which a recognition may pass from a simultaneous to a successive association. If a certain interval elapses before the elements of the earlier idea which gradually rise in consciousness, can produce a distinct feeling of familiarity, the whole process divides into two acts: into the *perception* and the *recognition*. The first is connected with the ordinary simultaneous assimilations only, while in the second the obscure, unassimilated elements of the earlier idea show their influence. The division between the parts is, accordingly, more distinct the greater the difference between the earlier impression and the new one. In such a case, not only is there usually a long period of noticeable inhibition between perception and recognition, but certain additional apperceptive processes, namely the processes of voluntary attention that take place in the state of recollection, also come to the aid of the association. As a special form of this kind of process we have the phenomenon called "mediate recognition". This consists in the recognition of an object, not through its own attributes, but through some accompanying mark or other, which stands in a chance connection with it, as, for example, when a person is recognized because of his companion. Between such a case and a case of immediate recognition there is no essential psychological difference. For even those characteristics that do not belong to the recognized object in itself, still belong to the whole complex of ideational elements that help in the preparation and final carrying out of the association. And yet, as we should naturally expect, the retardation which divides the whole recognition into two ideational processes, and often leads to the cooperation of voluntary recollection generally appears in its most evident form in mediate recognitions.

17. This simple process of recognition which takes place when we meet again an object that has been perceived once before, is a starting point for the development of various other associative processes, both those which like itself stand on the boundary between simultaneous and successive associations, and those in which the retardation in the form of assimilations and complications that leads to the success processes, is still more clearly marked. Thus, the recognition of an object that has often been perceived is easier and, therefore, as a rule an instantaneous process, which is also more like the ordinary assimilation because the feeling of familiarity is much less intense. *Sensible cognition* differs, generally but little from the recognition of single familiar objects. The logical distinction between the two concepts consist in the fact that recognition means the establishment of individual identity of the newly perceived with a formerly perceived object, while cognition is the subsumption of object under a familiar concept. Still, there is no real logical subsumption in a process of sensible cognition any more there is a fully developed class-concept under which the subsumption could be made. The psychological equivalent of such a subsumption is to be found in this case in the process of relating the impression in question to an indefinitely large number of objects. This presupposes an earlier perception of various objects which agree only in certain particular properties, so that the process of cognition approaches the ordinary assimilation more and more in its psychological character the more familiar the class to which the, perceived object belongs, and the more it agrees with the general characteristics of the class. In equal measure the feelings peculiar to the processes of cognition and recognition decrease and finally disappear entirely, so that when we meet very familiar objects we do not speak of a cognition at all. The process of cognition becomes evident only when the assimilation is *hindered* in some way, either because the perception of the class of objects in question has become unusual, or because the single object shows some unique characteristics. In such a case the simultaneous association may become successive by the separation of perception and cognition into two successive processes. Just in proportion as this happens, we have a specific *feeling of cognition* which is indeed related to the feeling of familiarity, but, as a result of the different conditions for the rise of the two, differs from it, especially in its temporal course.

b. Memory-processes.

18. Essentially different is the direction along which the simple process of recognition develops, when the hindrances to immediate assimilation which give rise to the transition from simultaneous to successive association are great enough, so that the ideational elements which do not agree with the new perception unite ~ either after the recognition has taken place or when there is no such recognition whatever ~ to form a special idea referred directly to an earlier impression. The process that arises under such circumstances is a memory-process and the idea that is perceived is a *memory-idea,* or *memory-image.*

18a. Memory-processes were the ones to which association-psychology generally limited the application of the concept association. But, as has been shown, these are associations that take place under especially complicated conditions. An understanding of the genesis of association was thus rendered impossible from the first, and it is easy to see that the doctrine accepted by the associationists is limited essentially to a logical rather than a psychological classification of the different kinds of association that are to be observed in memory-processes. A knowledge of these more complex processes is possible, however, only through a study starting with the simpler associative processes, for the ordinary simultaneous assimilations and simultaneous and successive recognitions present themselves very naturally as the antecedents of memory-associations. But even simultaneous recognition itself is nothing but an assimilation accompanied by a feeling which comes from the unassimilated ideational elements obscurely present in consciousness. In the second process these unassimilated elements serve to retard the process, so that the recognition develops into the primitive form of successive association. The impression is at first assimilated in the ordinary way, and then again in a second act with an accompanying feeling of recognition which serves to indicate the greater influence of certain reproduced elements. In this simple form of successive association the two successive ideas are referred to one and the same object, the only difference being that each time some different ideational and affective elements are apperceived. With *memory-associations* the case is essentially different. Here the heterogeneous elements of the earlier impressions predominate, and the first assimilation of the impression is followed by the formation of an idea made up of elements of the impression and also of those belonging, to earlier impressions, that are suitable for the assimilation because of certain of their components. The more the heterogeneous elements predominate, the more is the second idea different from the first, or, on the other hand, the more the like elements predominate, the more the two ideas will be *alike*. In any case the second idea is always a *reproduced* idea and distinct from the new impression as an independent compound.

19. The general conditions for the rise of memory-images may exhibit shades and differences which run parallel to the forms of recognition and cognition discussed above. Various modifications of the memory-processes may. arise from the different kinds of ordinary assimilation that we become acquainted with above, as the recognition of an object perceived once and that of an object familiar through *frequent* perceptions, and also from the cognition of a subject that is *familiar* in its general class-characteristics.

Simple recognition becomes a memory-process when the immediate assimilation of the impression is hindered by elements that belong not to the object itself, but to circumstances that attended its earlier perception. Just because the former perception occurred only once, or at least only once so far as the reproduction is concerned, these accompanying elements may be relatively clear and distinct and sharply distinguished from the surroundings of the new impression. In this way we have first of all transitional forms between recognition and remembering: the object is recognized, and at the same time referred to a particular earlier sense perception whose accompanying circumstances add a definite spacial and temporal relation to the memory-image. The memory-process is especially predominant in those cases where the element of the new impression that gave rise to the assimilation is entirely suppressed by the other components of the image, so that the associative relation between the memory-idea and the impression may remain completely unnoticed

19a. Such cases have been spoken of as "mediate memories", or "mediate associations". Still, just as with "mediate recognitions" we are, here too, dealing with processes that are fundamentally the same as ordinary associations. Take, for example, the case of a person who, sitting in his room at evening, suddenly remembers without any apparent reason a landscape that he passed through many years before; examination shows that there happened to be in the room a fragrant flower which he saw for the first time in that landscape. The difference between this and an ordinary memory-process in which the connection of the new impression with an earlier experience is clearly recognized, obviously consists in the fact that here the elements which recall the idea are pushed into the obscure background of consciousness other ideational elements. The not infrequent experience, commonly known as the "spontaneous rise" of ideas, in memory-image suddenly appears in our mind without any cause, is in all probability reducible in every case to such latent associations.

20. Memory-processes that develop from recognitions which have been *often* repeated and from *cognitions,* are in consequence of the greater complexity of their conditions, different from those connected with the recognition of objects perceived but once. When we perceive an object that is familiar either in its own individual characteristics or in those of its class, the range of possible associations is incomparably greater, and the way in which the memory-processes shall arise from a particular impression depend less on the single experiences that give rise to the association, than it does on the general disposition and momentary mood of consciousness and especially on the interference of certain active apperceptive processes and the intellectual feelings and emotions that are connected with them. When the conditions are so various, it is easy to see that as a general thing it is impossible, to calculate beforehand what the association will be. As soon as the act of memory is ended, however, the traces of its associative origin seldom escape careful examination, so that we are justified in regarding association as the universal and, only cause of memory-processes under all circumstances.

21. In thus deriving memory from association, it is not to be forgotten that every concrete memory-process is by no means a simple process, but is made up of a large number of elementary processes, as is apparent from the fact that it produced by a psychological development of its simple antecedents, namely, the simultaneous assassinations. The most important of these elementary processes is the assimilative interaction between some external impression and the elements of an earlier psychical compound, or between a memory-image already present and such elements. Connected with this there are two other processes that are characteristic for memory processes: one is the hindrance of the assimilation by unlike elements, the other the assassinations and complications connected with these elements and giving rise to a psychical compound which differs from the first impression and is referred more or less definitely to some previous experience, especially through its complications. This reference to the earlier experience shows itself through a characteristic feeling, the *feeling of*

remembering, which is related to the feeling of familiarity, but is in its temporal genesis characteristically different, probably in consequence of the greater number of obscure complications that accompany the appearance of the memory-image.

If we try to find the elementary processes to which both memory-processes and all complex associations are reducible, we shall find two kinds, combinations from *identity* and from *contiguity.* In general the first class is predominant when the process is more like an ordinary assimilation and recognition, while the second appears more prominently the more the processes approach mediate memory in character, that is, the more they take on the semblance of spontaneous ideas.

21 a. It is obvious that the usual classification, which makes all memory-processes associations by either similarity or contiguity, is entirely unsuitable if we attempt to apply it to the modes of psychological genesis that these processes manifest. On the other bland, it is too general and indefinite if we try to classify the processes logically according to their products, without reference to their genesis. In the latter case the various relations of subordination, superordination, and coordination, of cause and end, of temporal succession and existence, and the various kinds of spacial connection, find only inadequate expression in the very general concepts "similarity" and "contiguity". When, on the other hand, the manner of origin is studied, every memory-process is found to be made up of elementary processes that may be called partly associations by similarity, partly associations by contiguity. The assimilations which serve to introduce the process and also those which serve to bring about the reference to a particular earlier experience at its close, may be called associations by similarity. But the term "similarity" is not exactly suitable even here, because it is identical elementary processes that give rise to the assimilation, and when such an identity does not exist, it is always produced by the reciprocal assimilation. In fact, the concept of "association by similarity" is based on the presupposition that composite ideas are permanent psychical objects and that associations take place between these finished ideas. The concept itself must be rejected when once this presupposition is given up as entirely contradictory to psychical experience and fatal to a proper understanding of the same. When certain products of association, as, for example, two successive memory-images, are similar, this likeness is always reducible to processes of assimilation made up of elementary combinations through identity or contiguity. The association through identity may take place either between components that were originally the same, or between those that have gained this character through assimilation. Association by contiguity is the form of combination between those elements that hinder the assimilation, thus dividing the whole process into a succession of *two* processes, and also contributing to the memory-image those components which give it the character of an independent compound different from that of the impression which gave rise to it.

22. The character of *memory-ideas* is intimately connected with the complex nature of the memory-processes. The description of these ideas as weaker, but otherwise faithful, copies of the direct sensible idea, is as far out of the way as it could possibly be. Memory-images and direct sensible ideas differ not only in quality and intensity, but most emphatically in elementary composition. We may diminish the intensity of a

sensible impression as much as we like, but so long as it is perceptible at all it is an essentially different compound from a memory-idea. The *incompleteness* of the memory-idea is much more characteristic than the small intensity of its sensational elements. For example, when I remember an acquaintance, the image I have of his face and figure are not mere obscure reproductions of what I have in consciousness when I look directly at him, but most of the features do not exist at all in the reproduced ideas. Connected with the few ideational elements that are really present and that can be but little increased in number even when the attention is intentionally concentrated upon the task, are a series of combinations through contiguity and of complications, such as the environments in which I saw my acquaintance, his name, finally and more especially, certain affective elements that were present at the meeting. These accompanying components are what make the image a memory-image.

23. There are great individual differences in the effectiveness of these accompanying elements and in the distinctness of the sensational elements of the memory-image. Some persons locate their memory-images in space and time much more precisely than others do; the ability to remember colors and tones is also exceedingly different. Very few persons seem to have distinct memories for odors and tastes; in place of these we have, as substitute complications, accompanying motor sensations of the nose and taste-organs.

These various different functions connected with the processes of recognition and remembering are all included under the name "memory". This concept does not, of course, refer to any unitary psychical force, as faculty-psychology assumed, still, it is a useful supplementary concept in emphasizing the differences between different individuals. We speak of a faithful, comprehensive, and easy memory, or of a good spacial, temporal, and verbal memory, etc. These expressions serve to point out the different directions in which, according to the original disposition or habit of the person, the elementary assimilations and complications occur.

One important phenomenon among the various differences referred to, is the gradual *weakening of memory with old age.* The disturbances resulting from diseases of the brain agree in general with this phenomenon. Both are of special importance to psychology because they exhibit very clearly the influence of complications on memory-processes. One of the most striking symptoms of failing memory, in both normal and pathological cases, is the weakening of *verbal memory.* It generally appears as a lack of ability to remember, first. proper names, then names of concrete objects in the ordinary environments, still later abstract words, and finally particles that are entirely abstract in character. This succession corresponds exactly to the possibility of substituting in consciousness for single classes of words other ideas that are regularly connected with them through complication. This possibility it obviously greatest for proper names, and least for abstract particles, which can be retained only through their verbal signs.

17. APPERCEPTIVE COMBINATIONS

1. Associations in all their forms and also the closely related processes of fusion that give rise to psychical compounds, are regarded by us as passive experiences, because the feeling of activity, which is characteristic for all processes of volition and attention, never appears except in connection with the apperception of the *completed product,* presented as a *result* of the combination. Associations are, accordingly, processes that can arouse volitions, [p. 249] but are not themselves directly influenced by volitions. This is, however, the criterion of a *passive* process.

The case is essentially different with the second kind of combinations that are formed between different psychical compounds and their elements, the *apperceptive combinations.* Here the feeling of activity with its accompanying variable sensations of tension does not merely follow the combinations as an after-effect produced by them, but it precedes them so that the combinations themselves are *immediately recognized as formed with the aid of the attention.* In this sense they are called active experiences.

2. Apperceptive combinations include a large number of psychical processes that are distinguished in popular parlance under the general terms thinking, reflection, imagination, understanding, etc. These are all regarded as higher psychical processes than sense-perceptions or pure memory-processes, still, they axe all looked upon as different from one another. Especially is this true of the so-called functions of imagination and understanding. In contrast with this loose view of popular psychology and of the faculty-theory, which followed in its tracks, association-psychology sought to find a unitary principle by subsuming the apperceptive combinations of ideas also under the general concept of association, at the same time limiting the concept, as noted above, to successive association. This reduction to association was effected either by neglecting the essential subjective and objective distinguishing marks of apperceptive combinations, or by attempting to avoid the difficulties of an explanation, through the introduction of certain supplementary concepts taken from popular psychology. Thus, "interest" or "intelligence" was credited with an influence on associations. Very often this view was based on the erroneous notion that the recognition of certain distinguishing features in apperceptive combinations and associations meant the assertion of an absolute independence of the former from the latter. Of course, this is not true. All psychical processes are connected with associations as much as with the original sense-perceptions. Yet, just m associations always form a part of every sense-perception and in spite of that appear in memory-processes as relatively independent processes, so apperceptive combinations are based, entirely on associations, but their essential attributes are not traceable to these associations.

3. If we try to account for the essential attributes of apperceptive combinations, we may first of all divide the psychical processes that belong to this class into simple and *complex apperceptive functions.* The simple functions are those of *relating and* comparing, the complex those of *synthesis* and *analysis.*

A. SIMPLE APPERCEPTIVE FUNCTIONS

(Relating and Comparing.)

4. The most elementary apperceptive function is the *relating of two psychical contents to each other*. The grounds for such relating is always given in the single psychical compounds and their associations, but the actual *carrying out* of the process itself is a apperceptive activity through, which the *relation itself* assumes a special conscious content distinct from the contents which are related, though indeed inseparably connected with them. For example, when we recognize the identity of an object with one perceived before, or when we are conscious of a definite relation between a remembered event and a present impression, there is in both cases a relating apperceptive activity connected with the associations.

So long as the recognition remains a pure association, the process of relating is limited to the feeling of familiarity that follows the assimilation of the new impression either immediately or after a short interval. When, on the contrary, apperception is added to association, this feeling is supplied with a clearly recognized ideational substratum. The earlier perception and the new impression are separated in time and then brought into a relation of identity on the basis of their essential attributes. The case is similar when we are conscious of the motives of a memory-act. This also presupposes that a comparison of the memory-image with the impression that occasioned it, be added to the merely associative process which gave rise to the image. This, again, is a process that can be brought about only through active attention.

5. Thus, the *relating* function is brought into activity through associations, wherever they themselves or their products are made the objects of voluntary observation. This function is always connected, as the examples mentioned show, with the function of *comparing,* so that the two must be regarded as interdependent partial functions. Every act of relating includes a comparison of the related psychical contents, and a comparison is, in turn, possible only through the relating of the contents compared with one another. The only difference is that in many cases the comparison is completely subordinated to the end of reciprocally relating the contents, while in others it is in itself the end. We speak of a process of relating in the strict sense in the first case, and of a process of comparing in the second. I call it relating when I think of a present impression as the reason for remembering an earlier experience; I call it comparing, on the other hand, when I establish certain definite points of agreement or difference between the earlier and the present event.

6. *The process of comparing* is, in turn, made up of *two* elementary functions which are as a rule intimately interconnected: of the perception of *agreements* on the one hand, and of *differences* on the other. The erroneous view still frequent acceptance that the existence of psychical elements and compounds is the same as their apperceptive comparison. The two are to be held completely apart. Of course, there must be

agreements and differences in our psychical processes themselves, or we could not perceive them; still the comparing activity by which we perceive, is different from the agreements and differences themselves and additional to them.

7. Psychical elements, the sensations and simple feelings, are compared in regard to their agreements and differences and thus brought into definite systems, each of which contain such elements as are closely related. Within such a system, especially a sensational system, two kinds of comparison are possible: that in respect to quality, and that in respect to *intensity*. Then, too, a comparison between grades of *clearness is* possible when attention is paid to the way in which the elements appear in consciousness. In the same way comparison is applied to intensive and extensive psychical compounds. Every psychical element and every psychical compound, in so far as it is a member of a regular graded system, constitutes a *psychical quantity*. A determination of the value of such a quantity is possible only through its *comparison* with some other quantity of the same system. Quantity is, accordingly, an original attribute of every psychical element and compound. It is of various kinds, as intensity, quality, extensive (spacial and temporal) value, and, when the different states of consciousness are considered, clearness. But the *determination of quantity* can be effected only through the apperceptive function of comparison.

8. *Psychical* measurement differs from *physical* measurement in the fact that the latter may be carried out in acts of comparison separated almost indefinitely in time, because its objects are relatively constant. For example, we can determine the height of a certain mountain to-day with a barometer and then after a long time the height of another mountain and if no sensible changes in the configuration of the land have taken place in the interval, we can compare the results of our two measurements. Psychical compounds, on the other hand, are not relatively permanent objects, but continually changing processes, so that we can compare two such psychical quantities only under the condition that they axe presented in immediate succession. This condition has as its immediate corollaries: first, that there is no absolute standard for the comparison of psychical quantities, but every such comparison stands by itself and is of merely relative value; secondly, that finer comparisons are possible only between quantities of the same dimension, so that a transfer analogous to that by which the most widely separate physical quantities, such as periods of time and physical forces, are reduced to spacial quantities of one dimension, are out of the question in psychical comparisons.

9. It follows that not every relation between psychical quantities can be established by direct comparison, but this is possible only for certain particularly favorable relations. These favorable cases are 1) *the equality between two psychical quantities,* and 2) *the just noticeable difference between two such quantities,* as, for example, two sensational intensities of like quality, or two qualities of like intensity belonging to the same dimension. As a somewhat more complex case which still lies within the limits of immediate comparison we have 3) *the equality of two differences between quantities* especially when these quantities belong to contiguous parts of the same system. It is obvious that in each of these three kinds of psychical measurements the two fundamental functions in apperceptive comparison, the perception of agreements and

of differences, are both applied together. In the *first* case the second of two psychical quantities *A* and *B* is gradually varied until it agrees for immediate comparison with A. In the *second case A* and *B* are taken equal at first and then *B* is changed until it appears either just noticeably greater or just noticeably smaller than *A*. Finally, the *third* case is used to the greatest advantage when a whole line of psychical quantities, as, for example, of sensational intensifies, extending from *A* as a lower to *C* as an upper limit, is so divided by a middle quantity *B*, which has been found by gradual variations, that the partial distance *AB* is apperceived as equal to *BC*.

10. The most direct and most easily utilizable results derived from these methods of comparison are given by the *second* method, or the *method of minimal differences* as it is called. The difference between the Physical stimuli which corresponds to the just noticeable difference between psychical quantities is called the *difference-threshold of the stimulus*. The stimulus from which the resulting psychical process, for example, a sensation, can be just apperceived, is called the *stimulus-threshold.* Observation shows that the difference-threshold of the stimulus increases in proportion to the distance from the stimulus-threshold, in such a way that the *relation* between the difference-threshold and the absolute quantity of the stimulus, or the *relative differ threshold,* remain constant. If, for example, a certain sound whose intensity is 1 must be increased 1/3 in order that the sensation may, be just noticeably greater, one whose intensity is 2 must be increased 2/3, one 3, 3/3, etc., to reach the difference-threshold. This law is called *Weber's law,* after its discoverer E. H. Weber. It is easily understood when we look upon it as a law of apperceptive comparison. From this point of view it mull obviously be interpreted to mean that *psychical quantities are compared according to their relative values.*

This view that Weber's law is an expression of the *general law of the relativity of psychical quantities,* assumes that the psychical quantities that are compared, themselves increase in proportion to their stimuli within the limits of the validity of the law. It has not yet been possible to demonstrate the truth of this assumption on its physiological side, on account of the difficulties of measuring exactly the stimulation of nerves and sense-organs. Still, we have evidence in favor of it in the psychological experience that in certain special cases, where the conditions of observation lead very naturally to a comparison of absolute differences in quantity, the absolute difference threshold, instead of the relative threshold, is found to be constant. We have such a case, for example, in the comparison, within wide limits, of minimal differences in pitch. Then, too, in many cases where large differences in sensations are compared according to the third method described above, equal absolute stimuli-differences, not relative differences, are perceived as equal. This shows that apperceptive comparison follows two different principles under different conditions: a principle of *relative* comparison that finds its expression in Weber's law and must be regarded as the more general, and a principle of *absolute* comparison of differences which takes the place of the first under special conditions which favor such a form of apperception.

10 a. *Weber's law* has been shown to hold, first of all, for the intensity of sensations and then, within certain limits, for the comparison of *extensive* compounds, especially

temporal ideas, also, to some extent, for spacial ideas of sight and for motor ideas. On the other hand, it does not hold for the spacial ideas of external touch, obviously on account of the complexity of the local signs; and it can not be verified for sensational qualities. In fact, for the comparison of pitches the absolute, not the relative difference-threshold is constant within wide limits. Still, the scale of tonal intervals is relative, for every interval corresponds to a certain ratio between the number of vibrations (for example, an octave 1 : 2, a fifth 2 : 3, etc.). This is probably due to the relationship between clangs which is due to the relation of the fundamental tone to its overtones. Even where an absolute comparison takes place instead of a comparison according to Weber's law of relativity, we must not, of course, confuse this with the establishment of an absolute measure. That would presuppose an absolute unit, that is, the possibility of finding a constant standard, which, as noted above, is in the psychical world impossible. Absolute comparison must take the form of a *recognition of the equality of equal absolute difference.* This is possible in the various single cases without a constant unit. Thus, for example, we compare two sensational lines AB and BC according to their relative values, when we think in both cases of the relation of the upper to the lower extreme sensation. In such a case we judge *AB* and *BC* to be equal when $B/A = C/B$ (Weber's law). On the other hand, we compare *A B* and *B C* according to their absolute values when the difference between *C* and *B* in the single sensational dimension in question appears equal to that between *B* and *A*, that is, when $C - B = B - A$ (law of proportionality). Weber's law has sometimes been regarded as the expression of the functional relation between sensation and stimulus, and it has been assumed that the law holds for infinitely small changes on both sides. On this basis there has been given to it the mathematical form of the logarithmic function: sensation increases in proportion to the logarithm of the stimulus (Fechner's psycho-physical law).

The methods for the demonstration of Weber's law, of relations between psychical quantities, whether elementary compound, are usually called *psycho-physical methods.* is unsuitable, however, because the fact that physical here employed is not unique, but holds for all the experimental psychology. They could better be capable, for the measurement of psychical quantities". With these methods it is possible to follow one of *two* courses in relations mentioned as favorable for judgment. A direct mode of procedure is as follows: one of two psychical quantities *A* and *B*, as, for example, *A* is kept constant, and *B* is gradually varied until it stands in one of the relations mentioned, that is, either equals A or is just noticeably greater or smaller, etc. These are the *adjustment-methods.* Among these we have as the method most frequently applied and that which leads most directly to conclusions, the "method of minimal changes", and then as a kind of modification of this for the case of adjustment until equality is reached, the "method of average error". The second mode of procedure is to compare in a large number of cases any two stimuli, *A* and *B*, which are very little different, and to reckon from the number of cases in which the judgments are $A = B, A > B, A < B$, the position of the relations mentioned, especially the difference-threshold. These are the *reckoning-method.* The chief of these is the method known as that of "right and wrong cases". It would be more proper to call it the "method of three cases" (equality, positive difference, and negative difference). Details as to this and the other methods belong in a special treatise on experimental psychology.

There are two other interpretations of Weber's law still met with besides the psychological interpretation given above; they may be called the *physiological* and the *psycho-physical* theories. The first derives the law from hypothetically assumed relations in the conduction of excitations in the central nervous system. The second regards the law as a specific law of the "interaction between body and mind". The physiological interpretation is entirely hypothetical and in certain cases, as, for example, for temporal and spacial ideas, entirely inapplicable. The psycho-physical interpretation is based upon a view of the relation of mind which must be rejected by the psychology of to-day (cf. § 22, 8).

11. As special cases in the class of apperceptive comparisons generally falling under Weber's law we have the comparison of quantities that are the *relatively greatest sensational differences* or, when dealing with feelings, *opposites.* The phenomena that appear in such cases are usually gathered up in the class-name *contrasts.* In the department where contrasts have been most thoroughly investigated, in *residual sensations,* there is generally an utter lack of discrimination between two phenomena which are obviously entirely in origin, though the results are to a certain extent related. We may distinguish these a *physiological* and *psychological,* contrasts. Physiological contrasts are closely connected with. the phenomena of after-images, perhaps they are the same. *Psychological* contrasts are essentially different; they are usually pushed into the background by the stronger physiological contrasts when the impressions are more intense. They are distinguished from the physiological by two important characteristics. First, they do not reach their greatest intensity, when the brightness and saturation are greatest, but when they are at the medium stages, where the eye is most sensitive to changes in brightness and saturation. Secondly, they can be removed by comparison with an independent object. Especially the latter characteristic shows these contrasts to be unqualifiedly the products of comparisons. Thus, for example, when a grey square is laid on a black ground and close by a similar grey square is laid on a white ground and all is covered with transparent paper, the two squares appear entirely different; the one on the black ground looks bright nearly white, that on the white ground looks dark, nearly, black. Now after-images and irradiations are very weak when, the brightness of the objects is small, so that it may assumed that the phenomenon described is a, psychological contrast. If, again, a strip of black cardboard which is covered with the transparent piper., and therefore exactly the same grey as the two squares, is held in way that it connects the two squares the contrast is removed entirely, or, at least, very much diminished. If in this experiment a colored ground is used instead of the achromatic, the grey square will appear very clearly in the appropriate complementary color. But here, too, the contrast can be made to disappear through comparison with an independent grey object.

12. Psychical contrasts appear also in other spheres of sensation so far as the conditions for their demonstration are favorable. They are also especially marked in the case of feelings and may arise under proper conditions in the case of spacial and temporal ideas. Sensations of pitch are relatively most free, for most persons have a well developed ability to recognize absolute pitch and this tends to overcome contrast. In the case of *feelings* the effect of contrast is intimately connected with their general attribute

of developing toward certain opposites. Pleasurable feelings especially are intensified by unpleasant feelings immediately preceding, and the same holds for many feelings of relaxation following feelings of strain, as, for example, a feeling of fulfillment after expectation. The effect of contrast in the case of spacial and temporal ideas is most obvious when the same spacial or temporal interval is compared alternately with a longer and with a shorter interval. In the two cases the interval appears different, in comparison with the shorter it appears greatest in comparison with the longer, smaller. Here too the contrast between spacial ideas can be removed by bringing an object between the contrasted figures in such a way that it is possible easily to relate them both to it.

13. We may regard the phenomena that result from the apperception of impressions whose real character differs from that *expected,* as special modifications of psychical contrast. For example, we are prepared to lift a heavy weight, but in the actual lifting of the weight it proves to be lighter, or the reverse takes place and we lift a heavy weight instead of a light one as we expected: the result is that in the first case we underestimate, in the second overestimate the real weight. If a series of exactly equal weights of different sizes are made so that they look like a set of weights varying regularly from a lighter to a heavier, they will appear to be different in weight when raised. The smallest will seem to be the heaviest and the largest to be the lightest. The familiar association that. the greater volume is connected with the greater mass aids the contrast. The varying estimations of the weight, however, is the result of the contrast between the real and the expected sensation.

B. COMPLEX APPERCEPTIVE FUNCTIONS

(Synthesis and Analysis.)

14. When the simple processes of relating and comparing are repeated and combined several times, the complex psychical functions of *synthesis* and *analysis* arise. *Synthesis* is primarily the product of the *relating* activity of apperception, *analysis* of the *comparing* activity.

As a combining function *apperceptive synthesis* is based upon fusions and associations. It differs from the latter in the fact that some of the ideational and affective elements that are brought forward by the association are voluntarily emphasized and others are pushed into the background. The motives of the choice can be explained only from the whole previous development of the individual consciousness. As a result of this voluntary activity the product of this synthesis is a complex whole whose components all come from former sense-perceptions and associations, but in which the combination of these components usually varies more or less from the actual impressions and the combinations of these impressions that, are immediately presented in experience.

154

The ideational elements of a compound thus resulting, from apperceptive synthesis may be regarded as the substratum for the rest of its contents, and so we call such a compound in general an *aggregate idea*. When the combination of the elements is peculiar, that is, markedly different from the products of the fusion and associations, the aggregate idea and each of its relatively independent ideational components is called an *idea of imagination* or *image of imagination*. Since the voluntary synthesis of elements may vary more or less, according to the character of the motives that gave rise to it, from the combinations presented in sense-perception and association, it is obvious that practically no sharp line of demarcation can be drawn between images of imagination and those of memory. But we have a more essential mark of the apperceptive process in the positive characteristic of a voluntary synthesis than in the negative fact that the combination does not correspond in character to any particular sense-perception. This positive characteristic gives also the most striking external difference between images of imagination and those of memory. It consists in the fact that the sensational elements of an apperceptive compound are much more like those of an immediate sense-perception in clearness and distinctness, and generally in completeness and intensity. This is easily explained by the fact that the reciprocally inhibitory influences which the uncontrolled associations exercise on one another, and which prevent the formation of fixed memory-images, are diminished or removed by the voluntary emphasizing of certain particular ideational compounds. It is possible to mistake images of imagination for real experiences. With memory-images this is possible only when they become images of imagination, that is, when the memories are no longer allowed to arise passively, but are to -some extent produced by the will. Generally, too, there are voluntary modifications in them or a mixing of real with imagined elements. All our memories are therefore made up of "fancy and truth".[13] Memory-images change under the influence of our feelings and volition to images of imagination, and we generally deceive ourselves with their resemblance to real experiences.

15. From the aggregate ideas thus resulting from apperceptive synthesis there arise *two* forms of apperceptive activity in the opposite direction of *analysis*. The one is known in popular parlance as activity of the *imagination,* the second as activity of the *understanding*. The two are by no means different, as might be surmised from these names, but closely related and almost always connected with each other. Their fundamental determining motives are what distinguish them first of all and condition all their secondary differences as well as the reaction that they exercise on the synthetic function.

In the case of the activity of *"imagination"* the motive is the *reproduction of real experiences or of those analogous to reality*. This is the earlier form of apperceptive analysis and rises directly from associations. It begins with a more or less comprehensive aggregate idea made up of a variety of ideational and affective elements and embracing the general content of a complex experience in which the single

[13] The author [Wundt] remarks that the English word idea as here used corresponds to the German *Vorstellung*. Tr. [Judd]

components are only indefinitely distinguished. The aggregate idea is then divided in a series of successive acts into a number of more definite, connected compounds partly spacial, partly temporal in character. The primary voluntary synthesis is thus followed by analytic acts which may in turn give rise to the motives for a new synthesis and thus to a repetition of the whole process with a partially modified or more limited aggregate idea.

The activity of imagination shows *two* stages of development. The first is more *passive* and arises directly from the ordinary memory-function. It appears continually in the train of thought, especially in the form of an anticipation of the future, and plays an important part in psychical development as an antecedent of volitions. It may, however, in an analogous way, appear as a representation in thought of imaginary situations or of successions of external phenomena. The second, or *active,* stage of development is under the influence of a fixed idea of some end, and therefore presupposes a high degree of voluntary control over the images of imagination, and a strong interference, partly inhibitory, partly selective, with the memory-images that tend to push themselves into consciousness without voluntary action. Even the first synthesis of the aggregate idea is more systematic. An aggregate idea, when once formed, is held more firmly and subjected to a, more complete analysis into its parts. Very often these parts themselves are subordinate aggregate ideas to which the same process of analysis is again applied. In this way the principle of organic division according to the end in view governs all the products and processes of active imagination. The productions of *art* show this most clearly. Still, there are, in the ordinary play of imagination, the most various intermediate stages between passive imagination, or that which arises directly from memory, and active imagination, or that which is directed by fixed ends.

16. In contrast with this reproduction of real experiences or of such as may be thought of as real, which constitutes the content of the apperceptive functions that we include under the concept "imagination", the fundamental motive of the "understanding" is the *perception of agreements and differences and other derived logical relations between consent of experience.* Understanding also starts with aggregate ideas in which a number of experiences that are real or may he ideated as real, are voluntarily set in relation to one another and combined to a unitary whole. The analysis that takes place in this case, however, is turned by its fundamental motive in a different direction. It consists not merely in a clearer grasp of the single components of the aggregate idea, but in the establishment of the manifold relations in which these components stand to each other and which we may discover through comparison. As soon as such analyses have been made several times, results of the relating and comparing processes gained elsewhere can be employed in any particular case.

As a result of its more strict application of the elementary relating and comparing functions, the activity of understanding follows definite rules even in its external form, especially when it is highly developed. The principle that holds in general for imagination and even for mere remembering, that the relations of different psychical contents which are apperceived are presented, not simultaneously, but *successively,* so that in every case we pass on from one relation to a succeeding - this principle becomes for the activity of understanding, a rule of *discursive division of aggregate ideas.* It is expressed in the law of the duality *of the logical forms of thought,* according to which analysis resulting from relating comparison divides the content of the aggregate idea into *two* parts, subject and predicate, and may then separate each of these parts again once or several times. These second divisions give rise to grammatical forms that stand in a logical relation analogous to that of subject and predicate, such as noun and attributive, verb and object, verb and adverb. In this way the process of apperceptive analysis results in *judgment.*

For the psychological explanation of judgment it is of fundamental importance that it be regarded, not as a synthetic, but as an analytic function. The original aggregate ideas that are divided by judgment into their reciprocally related components, are exactly like ideas of imagination. The products of analysis that result are, on the other hand, not at in the case of imagination, images of more limited extent greater clearness, but *conceptual ideas,* that is ideas which stand, with regard to other partial ideas of the same whole, in some one of those relations which are discovered through the general relating and comparing functions. If we call the aggregate idea which is subjected to such a relating analysis a *thought,* then a *judgment* is a division of this thought into its components, and a *concept* is the product of such a division.

17. Concepts found in this way are arranged in certain general classes according to the character of the analyses that took place. These classes are the concepts of objects, attributes, and states. Judgment, as a division of the aggregate idea, sets an object in relation to its attributes or states, or various objects in relation to one another. Since a single concept can never, strictly speaking, be thought of by itself, but is always connected in the whole idea with one or more other concepts, the conceptual ideas are strikingly different from the ideas of imagination because of the indefiniteness and variableness of the former. This indefiniteness is essentially increased by the fact that a single concept may exist in an unlimited variety of modifications, since concepts which result from different cases of like judgment, may form components of many ideas that differ in their concrete characters. Such *general* concepts constitute, on account of the wide application of relating analysis to different contents of judgment, the great majority of all concepts; and they have a great number of corresponding single ideational contents. It becomes necessary, accordingly, to choose a single idea as a representative of the concept. This gives the conceptual idea a greater definiteness. At the same time there is always connected with this idea the consciousness that it is merely a representative. This consciousness generally takes the form of a characteristic feeling. This *conceptual feeling* may be traced to the fact that obscure ideas, which have the attributes that make them suitable to serve as representations of the concept, tend to force themselves into consciousness in the form of variable memory images. As evidence of this we have the fact that the feeling is very intense so long as any concrete image of the concept is chosen as its representative, as, for example, when a particular individual stands for the concept man, while it disappears almost entirely so soon as the representative idea differs entirely in content from the objects included under the concept. *Word-ideas* fulfill this condition and that is what gives them their importance as universal aids to thought. These aids are furnished to the individual consciousness in

a finished so that we must leave to social psychology the question of the psychological development of the processes of thought active in the formation of language.

18. From all that has been said it appears that the activities of imagination and understanding are not specifically different, but interrelated. inseparable in their rise and manifestations, and based at bottom on the same fundamental functions of apperceptive synthesis and analysis. What was true of the concept *"memory"* holds also of the concepts *"understanding"* and *"imagination"*: they are names, not of unitary forces or faculties, but of complex phenomena made up of elementary psychical processes of the usual, not of a specific, distinct kind. Just as memory is a general concept for certain associative processes, imagination and understanding are general concepts for particular forms of apperceptive activity. They have a certain practical value as ready means for the classification of an endless variety of differences in the capacity of various persons for intellectual activity. Each class thus found may in turn contain an endless variety of gradations and shades. Thus, neglecting the general differences in grade, we have as the chief forms of individual imagination the *perceptive* and the combining forms; as the chief form of understanding, the inductive and *deductive* forms, the first being mainly concerned with the single logical relations and their combinations, the second more with general concepts and their analysis. A person's talent is his total capacity relating from the special tendencies of both his imagination and understanding.

18. PSYCHICAL STATES

1. The normal state of consciousness upon which the discussion of the foregoing paragraphs has been based may undergo such a variety of changes that general psychology must give up the attempt to discuss them in detail. Then, too, the more important of these changes, namely, those which are observed in the various forms of nervous diseases, brain diseases, and insanity, belong to special branches of pathology which border upon psychology and are more or less dependent upon it. All that psychology can do is to indicate the main psychical conditions for such abnormal states of consciousness. We may distinguish in general, in accordance with what has been said about the attributes of psychical processes and their interconnection in consciousness *three* kinds of such conditions. They may consist 1) in the abnormal character of the psychical elements, 2) in the way psychical compounds are constituted, and 3) in the way psychical compounds are combined in consciousness. As a result of the intimate interconnection of these different factors it scarcely ever happens that one of these three conditions, each of which may appear in the most various concrete forms, is operative alone; but they usually unite. The abnormal character of the elements results in the abnormity of the compounds, and this in turn brings about changes in the general interconnection of conscious processes.

2.The psychical elements, sensations, and simple feelings, show only such changes as result from some disturbance in the normal relation between them. and their psycho-physical conditions. For sensations such changes may be reduced to an increase or decrease of the sensitivity for stimuli (by hyper-aesthesia, and anaesthesia) resulting especially from the of certain physiological influences in the sensory centers. The most important psychological symptom in this case is the *increased excitability* which is one of the most common components of complex psychical disturbances. In similar fashion,. changes in the simple feelings betray themselves in states of depression or exaltation as a decrease or increase in the affective excitability. These different states may be recognized from the way in which the emotions and volitional process occur. Thus, changes in the psychical elements can be demonstrated only by the influence that they exercise on the character of the various psychical compounds.

3. The defects in *ideational compounds* arising from peripheral or central anaesthesia are generally of limited importance. They have no far-reaching effect on the interconnection of psychical processes. It is essentially different with the relative *increase* in the intensity of sensations resulting from central hyperaesthesia. Its effect is especially important because under such circumstances reproduced sensational elements may become as intense as external sense-impressions. The result may be that a pure memory-image is objectified as a sense-perception. This is an *hallucination.* Or, when elements axe united which are partly from direct external: stimulation, partly from reproduction, the sense-impression may be essentially modified through the intensity of the reproduced elements. The result is then an *illusion of fancy.*[14]

The two are not always distinguishable. In many cases, to be sure, particular ideas can be shown to be illusions of fancy, but the presence of pure hallucinations is almost always doubtful because it is so easy to overlook some direct sensational elements. In fact, it is by no means improbable that the great majority of so-called hallucinations are illusions. These illusions are in their psychological character nothing but *assimilations.* They may be defined as assimilations in which the reproduced elements predominate. Just as normal assimilations are closely connected with successive associations, so for the same reason the illusions of fancy are closely related to the changes in the associative ideational processes to be discussed later (5).

4. In the case of *complex affective* and *volitional processes* the abnormal states of *depression* and *exaltation* are clearly distinguishable from the normal condition. The state of depression is due to the predominance of inhibitory, asthenic emotions, that of exaltation to a predominance of exciting, asthenic emotions, while at the same time we observe, in the first case a retardation or complete checking of resolution, in the second an exceedingly rapid, impulsive activity of the motive. In this sphere it is generally more difficult to draw the line between normal and abnormal conditions than in that of

---

[14] The expression "illusions of fancy" is used when this class of illusions is to be distinguished from the sense-illusions that appear in the normal state of consciousness, as, for example, the radiating for in of the stars, which is due to the refraction of light in the crystalline lens, or the varying apparent size of the sun or moon at the horizon and at the zenith.

ideational compounds, because even in normal mental life the affective states are continually changing. In pathological cases the change between states of depression and exaltation, which are often very striking, appear merely as an intensified oscillation of the feelings and emotions about an indifference-condition. States of depression and exaltation are especially characteristic symptoms of general psychical disturbances; their detailed discussion must therefore be left to psychical pathology. General psychical disturbances are always symptoms of diseases of the brain, so that these abnormities in affective and volitional processes are doubtless accompanied, like those of the sensations and ideas, by physiological changes. The nature of these changes is, however, still unknown. We can only surmise, in accordance with the more complex character of affective processes, either. that they are more extensive than the changes in central, excitability accompanying hallucinations and illusions, or that they effect the central cortical regions directly concerned in apperceptive processes.

5. Connected with these changes in the sensory excitability and with states of depression and exaltation, there are regularly simultaneous changes in the interconnection and course of psychical processes. Using the concept consciousness that we employ to express this interconnection, we may call these changes *abnormal changes of consciousness.* So long as the abnormity is limited to the single psychical compounds, ideas, emotions, and volitions, consciousness is of course changed because of the changes in its components, but we do no speak of an abnormality of consciousness itself until not merely the single compounds, but their combinations also exhibit some noticeable abnormities. These always arise, to be sure, when the elementary disturbances become greater, since the combination of elements to compounds and of compounds with one another are processes that pass continuously into each other.

Corresponding to the different kinds of combination that make up the interconnection of consciousness, there may be distinguished in general three kinds of abnormities of consciousness: 1) changes in the associations, 2) changes in the apperceptive combinations, and 3) changes in the relation of the two forms of combination to each other.

6. *Changes in associations* are the first to result directly from the elementary disturbances. The increase of sensory excitability changes normal assimilations into illusions of fancy, and this results in an essential disturbance in the associative processes of recognition: sometimes that which is known appears to be unknown, and then again what is unknown appears familiar, according as the reproduced elements are connected with definite earlier ideas, or are derived from perceptions that have only a remote relation to one another. Then, too, the increased sensory excitability tends to accelerate the association, so that the most superficial connections, occasioned by accidental impressions or by habit, are the ones that predominate. The states of depression and exaltation, on the other hand, determine mainly the quality and direction of the association.

160

In similar manner the elementary ideational and affective change influence *apperceptive* combinations, either retarding or accelerating them, or else determining their direction. Still, in these cases all marked abnormities in ideational or affective processes result in an increase, to a greater or less degree, of the difficulty of carrying out the processes connected with active attention, so that often only the simpler apperceptive combinations are possible, sometimes even only those which through practice have become simple associations. Connected with the last fact mentioned are the changes that take place in the relation between apperceptive and associative combinations. The influences discussed so far are in the main favorable to associations, but unfavorable to apperceptive combinations, and one of the most frequent symptoms of a far-reaching psychical abnormity is a great preponderance of associations. This is most obvious when the disturbance of consciousness is a continually increasing process, as it is in many cases of insanity. It is then observed that the functions of apperception upon which so-called imagination and understanding are based, are more and more supplanted by associations, until finally the latter are all that remains of the disturbance progresses still further, the associations gradually become more limited and confined to certain habitual combinations (fixed ideas). Finally this state gives place to one of complete mental paralysis.

7. Apart from mental diseases in the strict sense of the term the irregularities of consciousness just discussed are to be found in two conditions that appear in the course of normal life: in *dreams and hypnosis.*

The ideas of *dreams* come, at least to a great extent, from sensations, especially from those of the general sense, and are therefore mostly illusions of fancy, probably only, seldom pure memory-ideas that have become hallucinations. The decrease of apperceptive combinations in comparison with associations is also striking, and goes to explain the frequent modifications and exchanges of self-consciousness, the confusion of the judgment, etc. The characteristic of dreams that distinguishes them from other similar psychical states, is to be found, not so much in these. positive, as in their negative attributes. The increase of excitability which is attested by the hallucinations, is limited entirely to the *sensory* functions, while in ordinary sleep and dreams the external volitional activity is completely inhibited.

When the fanciful ideas of dreams are connected with corresponding volitional acts, we have the very infrequent phenomena of *sleep-walking,* which are related to certain forms of hypnosis. Motor concomitants are generally limited to articulations, and appear as talking in dreams.

8. *Hypnosis* is the name applied to certain states related to sleep and dreams and produced by means of certain definite psychical agencies. Consciousness is here generally in a condition halfway between waking and sleeping. The main cause of hypnosis is *suggestion,* that is, the communication of an idea strong in affective tone. This generally comes in the form of a command from some other person (outward suggestion), but may sometimes be produced by the subject himself, when it is called autosuggestion. The command or resolution to sleep, to make certain movements, to see

objects not present or not to see objects that are present, etc., - these are the most frequent suggestions. Monotonous stimuli, especially tactual stimuli are helpful auxiliaries. Then, too, there is a certain disposition of the nervous system of still unknown character, which is necessary for the rise of the hypnotic state and is increased when the state is repeatedly produced.

The first symptom of hypnosis is the more or less complete inhibition of volition, connected with a concentration of the attention on one thing, generally the commands of the hypnotizer (automatism). The subject not only sleeps at command, but retains in this state any position that is given him, however unnatural (hypnotic catalepsy). If the sleep becomes still deeper the subject carries out movements as directed, to all appearances automatically, and shows that ideas suggested to him appear like real objects (somnambulism). In this last state it is possible to give either motor or sensory suggestions to go into effect when the subject awakes, or even at some later time (terminal suggestions). The phenomena that accompany such "posthypnotic effects" render it probable that the latter are due either to a partial persistence of the hypnosis or (in the case of terminal suggestions) to a renewal of the hypnotic state.

9. It appears from all these phenomena that sleep and hypnosis are related states, differing only in that their mode of origin is different. They have as common characteristics the inhibition of volition, which permits only passive apperception, and a disposition toward aroused excitability in the sensory centers that brings about an assimilation of the sense-impressions which results in hallucinations. The characteristics that distinguish them are the complete inhibition of volition in sleep, especially of the motor functions, and the concentration in hypnosis of the passive attention on one thing. This concentration is conditioned by suggestion and is at the same time favorable to the reception of further suggestions. Still, these differences are not absolute, for in sleep-walking the will is not completely inhibited, while on the other hand it is inhibited in the first lethargic stages of hypnosis just as in ordinary sleep.

Sleep, dreams, and hypnosis are, accordingly, in all probability, essentially the same in their psychophysical conditions. These conditions are specially modified dispositions to sensational and volitional reactions, and can therefore, like all such dispositions, be explained on their physiological side only by assuming changes in the activity of certain central regions. These changes have not yet been investigation directly. Still, we may assume from the psychological symptoms that they consist in the inhibition of the activity in the regions connected with processes of volition and attention, and in the increase in the excitability of sensory centers.

9a. It is then, strictly speaking, a *physiological* problem to formulate a theory of sleep, dreams, and hypnosis. Apart from the general assumption based on psychological symptoms, of an inhibition of activity in certain parts of the cerebral cortex, and increase in the activity of other parts, we can apply only .one general neurological principle with any degree of probability. That is the principle of *compensation of functions,* according to which the inhibition of the activity of one region is always connected with an increase in the activity of the others interrelated with it. This interrelation may be either direct, *neurodynamic*, or indirect, *vasomotoric*. The first is probably due to the fact that energy which accumulates in one region as the result of inhibition, is discharged through the connecting fibers into other central regions. The second is due to contraction of the capillaries as a result of inhibition and a compensating dilation of the blood-vessels in other regions. The increased blood supply due to this dilation is in turn attended by an increase in the activity of the region in question.

Dreams and hypnosis are often made the subjects of mystical and fanciful hypotheses, in some cases even by psychologists. We hear of increased mental activity in dreams and of influence of mind on minds at a distance in dreams and hypnosis. Especially hypnotism has been used in modern times, in this way, to support superstitious spiritualistic ideas. In connection with "animal magnetism", which may be completely explained by the theory of hypnosis and suggestion, and in connection with "somnambulism", there are a great many cases of self-deception and intentional humbug. In reality all that can stand the light of thorough examination in these phenomena is in general readily explicable on psychological and physiological grounds; what is not explicable in this way has always proved on closer examination to be superstitious self-deception or intentional fraud.

CHAPTER 4

PSYCHICAL DEVELOPMENTS

19. PSYCHICAL ATTRIBUTES OF ANIMALS

1. The animal kingdom exhibits a series of mental developments which may be regarded as antecedents to the mental development of man. The mental life of animals shows itself to be in its elements and in the general laws of their combination everywhere the same as that of man.

Even the lowest animals (protozoa and coelenterata) manifest vital phenomena that allow us to infer ideational and volitional processes. They seize their food to all appearances spontaneously; they flee from pursuing enemies, etc. There are also to be found in the lowest stages of animal life traces of associations and reproductions and especially processes of sensible cognition and recognition. They reach a more advanced stage of development in higher animals only through the increase in the variety of ideas and in the length of time through which the memory-processes extend. From the like structure and development of the sense-organs we must draw the conclusion that the character of the sense-ideas are in general the same, the only difference being that in the lowest forms of life the sensory functions are limited to the general sense of touch, just as in the case of the higher organisms in the first stages of their individual development.

In contrast whith [sic] this uniformity of psychical elements and their simpler combinations there are great differences in all the processes connected with the development of *apperception*. *Passive* apperception is never absent as the basis for the simple impulsive acts that are found everywhere, but active apperception in the form of voluntary attention to certain impressions and a choice between different motives probably never exists except in the higher animals. Even here it is limited to the ideas and associations aroused by immediate sensible impressions, so that we can at most, if at all, only find the first beginnings of intellectual processes in the proper sense of the word, that is activities of imagination and understanding, even in the animals with the highest mental development. Connected with this fact is the other that higher animals have no developed language, though they are able to give expression to their emotions and even their ideas, when these are connected with emotions, through various expressive movements often related to those of man.

2. Though the development of animals is in general far behind that of man in spite of the qualitative likeness of the fundamental psychical processes, still, in two ways it is often superior. First, animals reach psychical maturity *much more rapidly*, and secondly, certain *single functions* particularly favored by the special conditions under which the species lives, are more highly developed. The fact of more rapid maturity is shown by the early age at which many animals, some immediately after birth, are able to receive relatively clear sense-impressions and to execute purposive movements. To be sure, there are very great differences among higher animals in this respect. For example, the chick just out of the shell begins to pick up grain, while the pup is blind at birth, and for a long time after clumsy in his movements. Yet, the development of the child seems to be the slowest and the most dependent on help and care from others.

3. The special *one-sided development of single functions* in some animals is still more striking. These functions show themselves in certain *impulsive acts* regularly connected with the satisfaction of certain needs, either of alimentation, reproduction, or protection, and in the development of the sense-perceptions and associations that form the motives for such acts. Such specially developed impulses are called *instincts.* The assumption that instincts belong only to animal and not to human consciousness is, of course, entirely unpsychological, and contradictory to experience. The disposition to manifest the general animal impulses, namely the alimentive and sexual impulses, is just as much a connate attribute of man as of the animals. The only thing that is characteristic is the special highly developed form of the purposive acts by which many animals reach the ends aimed at. Different animals, however, are very different in this respect. There are numerous lower and higher animals whose acts resulting from connate instincts show as few striking characteristics as those of men. It is also remarkable that domestication generally tends to do away with the instincts that animals had in their wild state, and to develop new ones that may generally be regarded as modifications of the wild instincts, as, for example, those of certain hunting dogs, especially those of bird-dogs and pointers. The relatively high development of certain special instincts in animals as compared with men, is simply a manifestation of the general unsymmetrical development of the former. The whole psychical life of animals consists almost entirely of the processes that are connected with the predominating instinct.

4. In general, instincts may be regarded as impulsive acts that arise from particular sensations and sense-feelings. The physiological sources of the sensations chiefly concerned in instincts are the *alimentary and genital organs.* All animal instincts may accordingly, be reduced to *alimentive* and sexual instincts, though in connection with the latter, especially in their more complex forms, there axe always auxiliary protective and social impulses which may 'be regarded, from the character of their origin, as special modifications of the sexual impulse. Among these auxiliary forms must be reckoned the impulses of many animals to build houses and nests, as in the case of beavers, birds, and numerous insects (for example, spiders, wasps, bees, ants), then, too, the instinct of animal marriage found chiefly among birds and appearing both in the monogamic and polygamic forms. Finally, the so-called "animal states", as those of the bees, of ants, and of termites, belong under this head. They are in reality not states, but

sexual communities, in which the social impulse that unites the individuals, as well as the common protective impulse, are modifications of the reproduction-impulse.

In the case of all instincts the particular impulsive acts arise from certain sense-stimuli partly external, partly internal. The acts themselves are to be classed as impulsive acts, or simple volitions, since they are preceded and accompanied by particular sensations and feelings that serve as simple motives. The complex, connate character of these acts can be explained only from general inherited attributes of the nervous system, as a result of which certain connate reflex mechanisms are immediately set in action by particular stimuli, without practice on the part of the individual. The purposive character of these mechanisms must also be regarded as a product of *general* psycho-physical development. As further evidence for this we have the fact that instincts show not only various individual modifications, but even a certain degree of higher development through individual practice. In this way, the bird gradually learns to build its nest better; bees accommodate their hive to changing needs; instead of sending out new colonies they enlarge the hive if they have the necessary room. Even abnormal habits may, be acquired by a single community of bees or ants; the first, for example, may learn to rob a neighboring hive instead of gathering the honey from the flowers, or the latter may acquire the remarkable habit of making the members of another species slaves, or of domesticating plant-lice for the sake of their honey. The rise, growth, and transmission of these habits as we can trace them, show clearly the way in which all complicated instincts may arise. Such an instinct never appears alone, but there are always *simpler.* forms of the same instinct in related classes and species. Thus the hole that the wall-wasp bores in the wall to lay her eggs in, is a primitive pattern of the ingenious hive of the honey-bee. Between these two extremes as the natural transition stage we have the hive of the ordinary wasp made of a few hexagonal cells constructed of cemented sticks and leaves.

We may, accordingly, explain the complex instincts as developed forms of originally simple impulses that have gradually differentiated more and more in the course of numberless generations, through the gradual accumulation of habits that have been acquired by individuals and then transmitted. Every single habit is to be regarded as a stage in this psychical development. Its gradual passage into a connate disposition is to be explained as a psycho-physical process of practice through which complex volitional acts gradually pass into purposive movements that follow immediately and reflexly [sic] the appropriate impression.

5. If we try to answer the general question of the *genetic relation of man to the animals* on the ground of a comparison of their psychical attributes, it must be admitted, in view of the likeness of psychical elements and of their simplest and most general forms of combination, that it is possible that human consciousness has developed from a lower form of animal consciousness. This assumption also finds strong support in the fact that the animal kingdom presents a whole series of different stages of psychical development and that every human individual passes through an analogous development. The doctrine of psychical development thus confirms in general the results of the theory of physical evolution, still we must not overlook the fact that the differences between the

psychical attributes of man and those of the animals, as expressed in the intellectual and effective processes resulting from apperceptive combinations, are much broader than the differences in their physical characteristics. Then, too, the great stability of the psychical condition of animals, which undergoes little change even in domestication, renders it exceedingly improbable that any of the present animal forms will develop much beyond the limits that they have already reached in their psychical attributes.

5a. The attempts to define the relation of man and animals from a psychological point of view vary between two extremes. One of these is the predominating view of the old psychology that the higher "faculties of mind", especially "reason", were entirely wanting in animals. The other is the wide-spread opinion of representatives of special animal psychology, that animals are fully equal to man in all respects, in ability to consider, to judge, to draw conclusions, in moral feelings, etc. With the rejection of faculty-psychology the first of these views becomes untenable. The second rests on the tendency prevalent in popular psychology to interpret all objective phenomena in terms of human thought, especially in terms of logical reflection. The closer psychological investigation of so-called manifestation of intelligence among animals shows, however, that they are in all cases fully explicable as simple sensible recognitions and associations, and that they lack the characteristics belonging to concepts proper and to logical operations. But associative processes pass without a break into apperceptive, and the beginnings of the latter, that is simple acts of active attention and choice, appear without any doubt in the case of higher animals, so that the difference is after all more one of the degree and complexity of the psychical processes than one of kind.

Animal instincts presented a very great difficulty to the older forms of psychology, such as the faculty-theory and the intellectualistic theories (§ 2). Since the attempt to deduce these instincts from the conditions given in each individual case led to an improbably high estimation of the psychical ability of the animal, especially when the instinct was more complex, the conclusion was often accepted that instincts are incomprehensible, or, what amounts to the same thing, due to connate ideas. This "enigma of the instincts" ceases to be an enigma when we come to look upon instincts, as we've done above, as special forms of impulsive action, and to consider them as analogous to the simple impulsive acts of men and animals, for which we have a psychological explanation. This is especially true when we follow the reduction of what were originally complicated acts, to impulsive or reflex movements in the phenomena of habit, so easily observed in the case of man, as, for example, the habituation to complex movements in learning to play the piano. It is often argued against this theory of instinct that it is impossible to prove empirically the transmission of acquired individual variations which we have assumed, that, for example, there are no certain observations in proof of the transmission of mutilations, as used to be asserted so frequently. Many biologists accept the view that all the properties of the organism arise through the selection resulting from the survival of the individual best adapted to natural conditions, that all such properties are accordingly deducible from "natural selection", and that in this way alone changes can be produced in the germ and transmitted to descendants. Though it must be admitted that an attribute acquired by a single individual, generally has no effect on the descendents, still, there is no apparent reason why habitual acts, which are indeed

167

indirectly due to outer natural conditions, but depend primarily on the inner psycho-physical attributes of the organism, may not cause changes in the nature of the germ when these acts are repeated through many Generations, just as well as the direct influences of natural selection. As further evidence for this view we have the fact that in some cases whole families inherit peculiar expressive movements or technical ability in some line. This does not exclude in any case the cooperation of natural influences, but is in full agreement with the facts of observation which show that these influences act in two ways: first, directly in the changes that natural selection brings about in the organism while the organism remains passive, and secondly, indirectly in the psycho-physical reactions that are caused by the outer influences, and then in turn give rise to changes in the organism. If we neglect the latter fact, we not only lose an important means of accounting for the eminently purposive character of animal organisms, but further, and more especially, we render impossible a psychological explanation of the gradual development of volition and its retrogradation into purposive reflexes as we see it in a large number of connate expressive movements (§ 20, 1).

20. PSYCHICAL DEVELOPMENT OF THE CHILD

1. The fact that the psychical development of man is regularly slower than that of most animals is to be seen in the much more gradual maturing of his *sense-functions*. The child, to be sure, reacts immediately after birth to all kinds of sense-stimuli, most clearly to impressions of touch and taste, with the least certainty to those of sound. Still, it is impossible to doubt that the special forms of the reaction-movements in all these cases are due to inherited reflexes. This is especially true for the child's crying when affected by cold and tactual impressions, and for the mimetic reflexes when he tastes sweet, sour, or bitter substances. It is probable that all these impressions are accompanied by obscure sensations and feelings, yet the character of the movements can not be explained from the feelings whose symptoms they may be considered to be, but must be referred to connate central reflex tracts.

Probably nothing is clear in consciousness until the end of the first month, and even then, as the rapid change of moods shows, sensations and feelings must be relatively very changeable. It is at about this time that we begin to observe symptoms of pleasurable and unpleasurable feelings in the child's laughter and in lively rhythmical movements of his arms and legs after certain impressions. Even the reflexes are not completely developed at first - a fact which we can easily understand when we learn from anatomy that many of the connecting fibers between the cerebral centers do not develop until after birth. Thus the associative reflex-movements of the two eyes are wanting. From the first each of the eyes by itself generally turns towards a light, but the movements of the two eyes are entirely irregular, and it is only in the course of the first three months that the normal coordination of the movements of the two eyes with a common fixation-point, begins to appear. Even then the developing regularity is not to be regarded as a result of complete visual perceptions, but, quite the reverse, as a symptom of the gradual functioning of a reflex-center, which then renders clear visual perceptions possible.

2. It is, generally speaking, impossible to gain any adequate information about the qualitative relations of psychical *elements* in the child's consciousness, for the reason that we have no certain objective symptoms. It is probable that the number of different tonal sensations, perhaps also the number of color-sensations, is very limited. The fact that children two years old not infrequently use the wrong names for colors ought not however, to be looked upon as unqualified evidence, that they do not have the sensation in question. It is much more probable that lack of attention and a confusion of the names is the real explanation in such cases.

Towards the end of the first year the *differential of feelings* and the related development of the various emotions take place, and show themselves strikingly in the characteristic expressive movements that gradually arise. We have unpleasurable feelings and joy, then in order, astonishment, expectation, anger, shame, envy, etc. Even in these cases the dispositions for the combined movements which express the single emotions, depend upon inherited physiological attributes of the nervous system, which generally do not begin to function until after the first few months, in a way analogous to the combined innervation of the ocular muscles. As further evidence of this we have the fact that not infrequently special peculiarities in the expressive movements are inherited by whole families.

3. The physical conditions for the rise of *spacial ideas* are connate in the form of inherited reflex-connections which make a relatively rapid development of these ideas possible. But for the child the spacial perceptions seem at first to be much more incomplete than they are in the case of many animals. There are manifestations of pain when the skin is stimulated, but no clear symptoms of localization. Distinct grasping movements develop gradually from the aimless movements that are observed even in the first days, but they do not, as a rule, become certain and consciously purposive until aided by visual perceptions, after the twelfth week. The turning of the eye toward a source of light as generally observed very early, is to be regarded as reflex. The same is true of the gradual coordination of ocular movements. Still it is probable that along with these reflexes there are developed spacial ideas, so that all we can observe is the gradual completion of these ideas from very crude beginnings, for the process is continuous and is always interconnected with its original physiological substratum. Even in the child the sense of sight shows itself to be decidedly more rapid in its development than the sense of touch, for the symptoms of visual localization are certainly observable earlier than those of tactual localization, and the grasping movements, as mentioned above, do not reach their full development until aided by the sense of sight. The field of *binocular* vision is much later in its development than that of monocular vision. The latter shows itself in the discrimination of directions in space. The beginnings of the development of a field for binocular vision coincide with the first coordination of ocular movements and belong, accordingly, to the second half of the first year. The perception of size, of distance, and of various three-dimensional figures remains for a long time very

169

imperfect. Especially, distant objects are all thought to be near at hand, so that they appear relatively small to the child.

4. *Temporal ideas* develop along with the spacial ideas. The ability to form regular temporal ideas and the agreeableness of these to the child shows itself in the first months in the movements of his limbs and especially in the tendency to accompany rhythms that are heard, with similar rhythmical movements. Some children can imitate correctly, even before, they can speak, the rhythmical melodies that they hear, in sounds and intonations. Still, the ideas of longer intervals are very imperfect even at the end of the first year and later, so that a child gives very irregular judgments as to the duration of different periods and also as to their sequence.

5. The development of *associations* and of *simple apperceptive combinations* goes hand in hand with that of spacial and temporal ideas. Symptoms of sensible recognition are observable from the very first days, in the rapidly acquired ability to find the mother's breast and in the obvious habituation to the objects and persons of the environment. Still, for a long time these associations cover only very short intervals of time, at first only hours, then days. Even in the third and fourth years children either forget entirely or remember only imperfectly persons who bay been absent for a few weeks.

The case with *attention* is similar. At first it is possible to concentrate it upon a single object only for a very short time, and it is obvious that passive apperception which always follows the predominating stimulus, that is the one whose affective tone is strongest, is the only form present. In the first weeks, however, a lasting attention begins to show itself in the way the child fixates and follows objects for a longer time, especially if they are moving; and at the same time we have the first trace of active apperception in the ability to turn voluntarily from one impression to another. From this point on, the ability becomes more and more fully developed; still, the attention, even in later childhood, fatigues more rapidly than in the case of adults, and requires a greater variety of objects or a more frequent pause for rest.

6. The development of *self-consciousness* keeps pace with that of the associations and apperceptions. In judging of this development we must guard against accepting as signs of self-consciousness any single symptoms, such as the child's discrimination of the parts of his body from objects of his environment, his use of the word "I", or even the recognition of his own image in the mirror. The adult savage who has never seen his own reflected image before, takes it for some other person. The use of the personal pronoun is due to the child's imitation of the examples of those about him. This imitation comes at very different times in the cases of different children, even when their intellectual development in other respects is the same. It is, to be sure, a symptom of the presence of self-consciousness, but the first beginnings of self-consciousness may have preceded this discrimination in speech by a longer or shorter period of time as the case may be. Again, the discrimination of the body from other objects is a symptom of exactly the same kind. The re cognition of the body is a process that regularly precedes that of the recognition of the image in the mirror, but one is as little a criterion of the beginning of self-consciousness as the other. They both presuppose the existence of

some degree of self-consciousness beforehand. Just as the developed self-consciousness is based upon a number of different conditions, so in the same way the self-consciousness of the child is from the first a product of several components, partly ideational in character, partly affective and volitional. Under the first head we have the discrimination of a *constant* group of ideas, under the second the development of certain interconnected processes of attention and volitional acts. The constant group of ideas does not necessarily include all parts of the body, as, for example, the legs, which are usually covered, and it may, as is more often the case, include external objects, as, for example, the clothes generally worn. The subjective affective and volitional components, and the relations that exist between these and the ideational components in external volitional acts, are the factors that exercise the decisive influence. Their greater influence is shown especially by the fact that strong feelings, especially those of pain, very often mark in an individual's memory the first moment to which the continuity of his self-consciousness reaches back. But there can be no doubt that a form of self-consciousness, even though less interconnected, exists even before this first clearly remembered moment, which generally comes in the fifth or sixth year. Still, since the objective observation of the child is not supplied at first with any certain criteria, it is impossible to determine the exact moment when self-consciousness begins. Probably the traces of it begin to appear in the first weeks; after this it continually becomes clearer under the constant influence of the conditions mentioned, and increases in temporal extent just as consciousness in general does.

7. The development of *will* is intimately connected with that of self-consciousness. It may be inferred partly from the development of attention described above, partly from the rise and gradual perfection of *external volitional acts,* whose influence on self-consciousness has just been mentioned. The immediate relation of attention to will appears in the fact that symptoms of active attention and voluntary action come at exactly the same time. Very many animals execute immediately after birth fairly perfect impulsive movements, that is, simple volitional acts. These are rendered possible by inherited reflex-mechanisms of a complex character. The new-born child, on the contrary, does not show any traces of such impulsive acts. Still, we observe in the first days the earliest beginnings of simple volitional acts of an impulsive character, as a result of the reflexes caused by sensations of hunger and by the sense-perceptions connected with appeasing it. These are to be seen in the evident quest after the sources of nourishment. With the obvious growth of attention come the volitional acts connected with impressions of sight and hearing: the child purposely, no longer merely in a reflex way, follows visual objects, and turns his head towards the noises that he hears. Much later come the movements of the outer muscles of the limbs and trunk. These, especially the muscles of the limbs, show from the first lively movements, generally repeated time and time again. These movements are accompanied by all possible feelings and emotions, and when the latter become differentiated, the movements begin gradually to exhibit certain differences characteristic for the quality of the emotions. The chief difference consists in the fact that rhythmical movement accompany pleasurable emotions, while arrhythmical and, as rule, violent movements result when the emotions are unpleasurable. These expressive movements, which must be looked upon as reflexes attended by feelings, then, as soon as the attention begins to

turn upon the surroundings, pass as occasion offers into ordinary *voluntary* expressive movements. Thus, the child shows through the different accompanying symptoms that he not only feels pain, annoyance, anger, etc., but that the wishes to give expression to these emotions. The first movements, however, in which an antecedent motive is to be recognized beyond a doubt, are the *graying movements* which begin in the twelfth to the fourteenth week. Especially at first, the foot takes part in these movements as well as the hand. We have here also the first clear symptoms of sense-perception, as well as the first indications of the existence of a simple volitional process made up of motive, decision, and act. Somewhat later intentional imitative movements are to be observed. Simple mimetic imitations, such as puckering the lips and frowning, come first, and then pantomimetic, such as doubling up the fist, beating time, etc. Very gradually, as a rule not until after the beginning of the second half of the first year, *complex* volitional acts develop from these simple ones. The oscillation of decision, the voluntary suppression of an intended act or one already begun, commence to be clearly observable at this period.

Learning to walk, which usually begins in the last third of the first year, is an important factor in the development of voluntary acts in the proper sense of the term. Its importance is due to the fact that the going to certain particular places furnishes the occasion for the rise of a number of conflicting motives. The learning itself is to be regarded as a process in which the development of the will and the effect of inherited dispositions to certain particular combinations of movements are continually interacting upon each other. The first impulse for the movement comes from volitional motives; the purposive way in which it is carried out, however, is primarily an effect of the central mechanism of coordination, which in turn is rendered continually more and more purposive as a result of the individual's practice directed by his will.

8. The development of the child's *ability to speak* follows that of his other volitional acts. This, too, depends on the cooperation of inherited modifications in the central organ of the nervous system on one hand, and outside influences on the other. The most important outside influences in this case are those that come from the speech of those about the child. In this respect the development of speech corresponds entirely to that of the other expressive movements, among which it is, from its general psycho-physical character, to be classed. The earliest articulations of the vocal organs appear as reflex phenomena, especially accompanying pleasurable feelings and emotions, as early as the second month. After that they increase in variety and exhibit more and more the tendency to repetition (for example, ba-ba-ba, da-da-da, etc.). These expressive sounds differ from those of many animals only in their greater and continually changing variety. They are produced on all possible occasions and without any intention of communicating anything, so that they are by no means to be classed as elements of speech. Through the influence of those about the child these sounds generally become elements of speech after the beginning of the second year. This result is brought about chiefly by certain imitative movements. It comes, in the form of sound-sensations, from two sides. On the one hand, the child imitates adults, on the other, adults imitate the child. In fact, as a rule, it is the adults who begin the imitating; they repeat the involuntary articulations of the child and attach a particular meaning to them, as, for

example, "pa-pa" for father, "ma-ma" for mother, etc. It is not until later, after the child has learned to use these, sounds in a particular sense though intentional imitation, that he repeats other words of the adults' language also, and even then he modifies them to fit the stock of sounds that he is able to articulate.

Gestures are important as means by which adults, more instinctively than voluntarily, help the child to understand the words they use. These are generally indicative gestures towards the objects; less frequently, ordinarily only in the case of words meaning seine activity such as strike, cut, walk, sleep, etc., they take the form of depicting gestures. The child has a natural understanding for these gestures, but not for words. Even the onomatopoetic words of child-speech (such as bow-bow for dog, etc.) never become intelligible to him until the objects have been frequently pointed out. The child is not the creator of these words, but it is rather the adult who seeks instinctively to accommodate himself in this respect also to the stage of the child's consciousness.

All this goes to show that the child's learning to speak is the result of a series of associations and apperceptions in whose formation both the child and those about to take part. Adults voluntarily designate particular ideas with certain words taken from the expressive sounds made by the child, or with onomatopoetic words made arbitrarily after the pattern of the first class. The child apperceives this combination of word and idea after it has been made intelligible to him with gestures, and associates it with his own imitative articulative movements. Following the pattern of these first apperceptions and associations the child their forms others, by imitating of his own accord more and more the words and verbal combinations that he accidentally hears adults using, and by making the appropriate associations with their meanings. The whole process is thus the result of a psychical interaction between the child and those about him. The sounds are at first produced by the child alone, those about him take up these sounds and make use of them for purposes of speech.

9. The final development that comes from all the simpler processes thus far discussed, is that of the *complex function of apperception,* that is the relating and comparing activities, and the activities of imagination and understanding made up of these (§ 17).

Apperceptive combination in its first form is exclusively the *activity of imagination,* that is the combination, analysis, and relating of concrete sensible ideas. Thus, individual development corroborates what has been said in general about the genetic relation of these functions. On the basis of the continually increasing association of immediate impressions with earlier ideas, there arises in the child, as soon as his active attention is aroused, a tendency to form such combinations voluntarily. The number of memory-elements freely combining with the impression and added to it, furnish us with a measure for the fertility of the individual child's imagination. As soon as this combining activity of imagination has once begun to operate, it shows itself with an impulsive force that the child is unable to resist, for there is not as yet, as ill the case of adults, any activity of the understanding to prescribe definite intellectual ends regulating and inhibiting the free sweep of the ideas of imagination.

This unchecked relating and coupling of ideas in imagination is connected with volitional impulses aiming to find for the ideas some starting-points in immediate sense-perception, however vague these starting-points may be. This is what gives rise to the child's *play-impulse.* The earliest games of the child are those of pure imagination; while, on the contrary, those of adults (cards, chess, lotto, etc.) are almost as exclusively intellectual games. Only where aesthetical demands exert an influence are the games of adults the productions of the imagination (drama, piano-playing, etc.), but even here they are not wholly untrammeled like those of the child, but are regulated by the understanding. When the play of a child takes its natural course, it shows at different periods of its development all the intermediate stages between the game of pure imagination and that in which imagination and understanding are united. In the first years this play consists in the production of rhythmical movements of the arms and legs, then the movements are carried over to external objects as well, with preference to such objects as give rise to auditory sensations, or such as are of bright colors. In their origin these movements are obviously impulsive acts aroused by certain sensational stimuli and dependent for their purposive coordination on inherited traits of the central nervous organs. The rhythmical order of the movements and of the feelings and sound-impressions produced by them, obviously arouse pleasurable feelings, and this very soon results in the voluntary repetition of the movements. After this, during the first years, play becomes gradually a voluntary imitation of the occupations and scenes that the child sees about him. The range of imitation then widens and is no longer limited to what is seen, but includes a free reproduction of what is heard in narratives. At the same time the interconnection between ideas and acts begins to follow a more fixed plan. This is the regulative influence of the activity of understanding, which shows itself in the games of later childhood in prescribed rules. This development is often accelerated through the influence of those about the child and through artificial forms of play generally invented by adults and not always suited to the child's imagination; still, the development is to be recognized as natural and necessarily conditioned by the reciprocal interconnection of associative and apperceptive processes, since it agrees with the general development of the intellectual functions. The way in which the processes of imagination are gradually curtailed and the functions of understanding more and more employed, renders it probable that the curtailing is due not so much to a quantitative decrease of imagination as to an obstruction of its action through abstract thinking. When this has once set in, because of the predominating exercise of abstract thinking, the activity of imagination may itself through lack of use be interfered with. This view seems to be supported by the fact that savages usually have all through their lives an imaginative play-impulse related to that of the child.

10. From imaginative forms of thought as a starting point the *functions of understanding* develop very gradually in the way already described. Aggregate ideas that are presented in sense-perception or formed by the combination, activity of imagination are divided into their *conceptual* components, into objects and their attributes, into objects and their activities, or into the relations of different objects to one another. The decisive symptom for the rise of the functions of understanding is therefore the formation of *concepts.* On the other hand, actions that can be explained from the point of view of the observer by logical reflection, are by no means proofs of

the existence of such reflection on the part of the actor, for they are very often obviously derived from associations, just as in the case of animals. In the same way there may be the first beginnings of speech without abstract thinking in any proper sense, since words refer originally only to concrete sensible impressions. Still, the more perfect use of language is not possible until ideas are conceptually analyzed, related, and transferred, even though the processes are in each case entirely concrete and sensible. The development of the functions of understanding and that of speech accordingly go hand in hand, and the latter is an indispensable aid in retaining concepts and fixing the operations of thought.

10a. Child-psychology often suffers from the same mistake that is made in animal psychology: namely, that the observations aren't interpreted objectively, but are filled out with subjective reflections. Thus, the earliest ideational combinations, which are in reality purely associative, are regarded as acts of logical reflection, and the earliest mimetic expressive movements, as, for example, those of a new-born child due to taste-stimuli, are looked upon as reactions to feelings, while they are obviously at first nothing but connate reflexes which may, indeed, be accompanied by obscure concomitant feelings, but even these can not be demonstrated with certainty. The ordinary view as to the development of volition and of speech, labors under a like misconception. Generally there is a tendency to consider the child's language, because of its peculiarities, as a creation of his own. Closer observation, however, shows that it is created by those about him, though in doing this they use the sounds that the child himself produces, and conform as far as possible to big stage of consciousness. Thus it comes that some of the very detailed and praise-worthy accounts of the mental development of the child in modern literature can serve only as sources for finding objective facts. Because they stand on the basis of a reflective popular psychology, their psychological deductions require correction along the lines marked out above.

21. DEVELOPMENT OF MENTAL COMMUNITIES

1. Just as the psychical development of the child is the resultant of his interaction with his environment, so matured consciousness stands continually in relation to the mental community in which it has a receptive and an active part. Among most animals such a community is entirely wanting. In animal marriage, animal states, and flocks, we have only incomplete forerunners of mental communities, and they are generally limited to the accomplishment of certain single ends. The more lasting forms, animal marriage and the falsely named animal states, are really sexual communities the more transient forms or flocks, as, for example, flocks of migratory birds, are communities for protection. In all these cases it is certain instincts that have grown more and more fixed through transmission, which hold the individuals together. The community, therefore, shows the same constancy as instinct in general, and is very little modified by the influences of individuals.

While animal communities are, thus, mere enlargements of the individual existence, aiming at certain physical vital ends, human development seeks from the first so to unite the individual with his mental environment that the whole is capable of development, serving at once the satisfaction of the physical needs of life and the pursuit of the most various mental ends, while permitting also great variations in these ends. As a result the forms of human society are exceedingly variable. The more fully developed forms, however, enter into a continuous train of historical development which extends the mental ties that connect individuals almost unlimitedly beyond the bounds of immediate spacial and temporal proximity. The final result of this development is the formation of the notion of *humanity* as a great general mental community which is divided up according to the special conditions of life into single concrete communities, peoples, states, civilized societies of various kinds, races, and families. The mental community to which the individual belongs is, therefore, not one, but a changing plurality of mental unions which are interlaced in the most manifold ways and become more and more numerous as development progresses.

2. The problem of tracing these developments in their concrete forms or even in their general interconnection. belongs to the history of civilization and to general history, not to psychology. Still, we must give some account here of the general psychical conditions and the psychical processes arising from these conditions that distinguish social from individual life.

The condition which is prime necessity of every mental community at its beginning, and a continually operative factor in its further development, is the *function of speech*. This is what makes the development of mental communities from individual existences psychologically possible. In its origin it comes from the expressive movements of the individual, but as a result of its development it becomes the indispensable form for all the common mental contents. These common contents, or the mental processes which belong to the whole community, may be divided into *two* classes, which are merely interrelated components of social life, not distinct processes any more than are the processes of ideation and volition in individual experience. The first of these classes is that of the *mythological ideas,* where we find especially the accepted conclusions on the question of the content and significance of the world - these are the *mythological ideas.* The second class consists of the *common motives of volition,* which correspond to the common ideas and their attending feelings and emotions - these are the *laws of custom.*

A. SPEECH

3. We obtain no information in regard to the *general development of speech* from the individual development of the child, because here the larger part of the process depends on those about him rather than on himself. Still, the fact that the child learns to speak at all, shows that he has psychical and physical traits favorable to the reception of language when it is communicated. In fact, it may be assumed that these traits would,

176

even if there were no communications from without, lead to the development of some kind of expressive movements accompanied by sounds, which would form an incomplete language. This supposition is justified 'by observations on the deaf and dumb, especially deaf and dumb children who have grown up without any systematic education. In spite of this lack of education, an energetic mental intercourse may take place between them. In such a case, however, since the deaf and dumb can perceive only *visual* signs, the intercourse must depend on the development of a natural *gesture-language* made up of a combination of significant expressive movements. Feelings are in general expressed by mimetic movements, ideas by pantomimetic, either by pointing at the object with the finger or by drawing some kind of picture of the idea in the air, that is, by means of *indicative* or *depicting* gestures. There may even be a combination of such signs corresponding to a series of successive ideas, and thus a kind of sentence may be formed, by means of which things are described and occurrences narrated. This natural gesture-language can never go any further, however, than the communication of concrete sensible ideas and their interconnection. Signs for abstract concepts are entirely wanting.

4. The primitive development of articulate language can hardly be thought of except after the analogy of the rise of this natural gesture-language. The only difference is that in this case the ability to hear results in the addition of a third form of movements to the mimetic and pantomimetic movements. These are the *articulatory movements,* and since they are much more easily perceived, and capable of incomparably more various modification, it must of necessity follow that they, soon exceed the others in importance. But just as mimetic and pantomimetic gestures owe their intelligibility to the immediate relation that exists between the character of the movement and its meaning,-so here also we must presuppose a like relation between the original articulatory movement and its meaning. Then, too,, it is not improbable that articulation was at first aided by accompanying mimetic and pantomimetic gestures. As evidence for this view we have the unrestrained use of such gestures by savages, and the important part they play in the child's learning to speak.. The development of articulate language is, accordingly, in all probability to be thought of as a process of differentiation, in which the articulatory movements have gradually gained the permanent ascendancy over a number of different variable expressive movements that originally attended them, and have dispensed with these auxiliary movements as they themselves gained a sufficient degree of fixity. Psychologically the process may be divided into *two* acts. The first consists in the expressive movements of the individual member of the community. These are impulsive volitional acts, among which the movements of the vocal organs gain the ascendancy over the others in the effort of the individual to communicate with his fellows. The second consists in the subsequent associations between sound and idea, which gradually become more fixed, and spread from the centers where they originated through wider circles of society.

5. From the first there are other physical and psychical conditions that take part in the formation of language and produce continual and unceasing modifications in its components. Such modifications may be divided into two classes: those of *sound* and those of *meaning.*

The first class have their physiological cause in the gradual changes that take place in the physical structure of the vocal organs. These changes are, to a great extent at least, either physically or psycho-physically conditioned. They come partly from the general changes which the transition from a savage to a civilized condition produces in the physical organism, partly from the special conditions that result from increased practice in the execution of articulatory movements. Many phenomena go to show that the gradually increasing rapidity of articulation is of especially great influence. Then, too, the words that are in any way analogous effect one another in a way that indicates the interference of the psychical factor of association.

As the change in sound modifies the outer form of words, so the change in meaning modifies their inner content. The original association between a word and the idea it expresses is modified by the substitution of another different idea. This process of substitution may be several times repeated with the same word. The change in the meaning of words depends, therefore, on a gradual modification of the associative conditions determining the ideational complication that shall arise in the fixation-point of consciousness when a word is heard or spoken. It may, accordingly, be briefly defined as a shifting of the ideational component of the complications connected with articulate sounds.

These changes in the sound and meaning of words operate together in bringing about the gradual disappearance of the originally necessary relation between sound and meaning, so that a word finally- comes to be looked upon as a mere external sign of the idea. This process is so complete that even those verbal forms in which this relation seems to be still retained, onomatopoetic words, appear to be, for the most part, products of a relatively late and secondary assimilative process which seeks to reestablish the lost affinity between sound and meaning.

Another important consequence of this combined action of changes in sound and meaning, is to be found in the fact that many words gradually lose entirely their original concrete sensible significance, and become signs of general concepts and means for the expression of the apperceptive relating and comparing functions and their products. In this way *abstract thinking* is developed. It would be impossible without the change in meaning of words upon which it is based and it is, therefore, a product of the psychical and psycho-physical interactions from which the progressive development of language results.

6. Just as the components of language, or words, are undergoing a continual development in sound and meaning, so in the same way, though generally more slowly, changes are going on in the combinations of these components into complete wholes, that is, in *sentences*. No language can be thought of without some such syntactic order of its words. Sentences and words are, therefore, equally primitive as psychological forms of thought. In a certain sense the sentence may even be called the earlier, for, especially in the more incomplete stages of language, the words of a sentence are so uncertainly distinguished that they seem to be nothing but the products of a breaking up of an originally unitary thought expressed by the whole sentence. There is no

universal rule for the order of words, any more than there is for the relation of sound to meaning. The order that logic favors with a view to the relations of reciprocal logical dependence between concepts, has no psychological universality; it appears, in fact, to be a fairly late product of development, due in part to arbitrary convention, and approached only by the prose forms of some modern languages which are syntactically nearly fixed. The original principle followed in apperceptive combination of words is obviously this, *the order of the words corresponds to the succession of ideas.* Especially those parts of speech that represent the ideas which arouse the most intense feelings and attract the attention, are placed first. Following this principle, certain regularities in the order of words are developed in any given community. In fact, such a regularity is to be observed even in the natural gesture-language of the deaf and dumb. Still, it is easy to understand that the most various modifications in this respect may appear under special circumstances, and that the possible range of these modifications is very great. In general, however, the habits of association lead more and more to the fixation of particular syntactic forms, so that a certain rigidity usually results.

Apart from the general laws presented in the discussion of apperceptive combinations, and there shown to arise from the general psychical functions of relating and comparing, the detailed discussion of the characteristics of syntactic combinations and their gradual changes, must be left, in spite of their psychological importance, to social psychology, because they depend so much on the specific dispositions and conditions of civilization in a given community.

B. MYTHS

7. The development of *myths* is closely related to that of language. Mythological thought is based, to be sure, just as language itself, upon certain attributes that are never lost in human consciousness; still, these attributes are modified and limited by a great variety of influences. As the fundamental function which in its various forms of activity gives rise to all mythological ideas, we have a characteristic kind of apperception belonging to all naive consciousness and suitably designated by the name personifying apperception. It consists in the complete determination of the apperceived objects through the nature of the perceiving subject. The subject not only sees his own sensations, emotions, an voluntary movements reproduced in the objects, but even his momentary affective state is in each case especially influential in determining this view of the phenomena perceived, and in arousing ideas of their relations to his own existence. As a necessary result of such a view the same *personal* attributes that the subject finds in himself are assigned to the object. The *inner* attributes, of feeling, emotion, etc., are never omitted, while the outer attributes of voluntary action and other manifestations like those of men, are generally dependent on movements actually perceived. The savage may thus attribute to stones, plants, and works of art, an inner capacity for sensations and feelings and their resulting effects, but he usually assumes immediate action only in the case of moving objects, such as clouds, heavenly bodies,

winds, etc. In all these cases the personification is favored by associative assimilations which may readily reach the intensity of illusions of fancy.

8. Myth-making, or personifying, apperception is not to be regarded as a special form or even as a distinct sub-form of apperception. It is nothing but the natural incentive stage of apperception in general. The child shows continually obvious traces of it, partly in the activities of his imagination in play, partly in the fact that strong emotions, especially fear and fright, easily arouse illusions of fancy with an affective character analogous to that of the emotion. In this case, however, the manifestations of a tendency to form myths are early checked and soon entirely suppressed through the influences of the child's environment and education. With savage and partly civilized peoples it is different. There the surrounding influences present a whole mass of mythological ideas to the individual consciousness. These, too, originated in the minds of individuals, and have gradually become fixed in some particular community, and in continual interrelation with language have, like the latter, been transmitted from generation to generation and become gradually modified in the transition from savage to civilized conditions.

9. The direction in which these modifications take place, is determined in general by the fact that the affective state of the subject at the time is, as above remarked, the chief. influence in settling the character of the myth-making apperception. In order to gain some notion of the way in which the affective state of the subject has changed from the first beginnings of mental development to the present, we must appeal to the history of the development of mythological ideas, for other evidences are entirely wanting. It appears that in all cases the earliest mythological ideas referred, on the one hand, to the personal fate in the immediate future, and were determined, on the other, by the emotions aroused by the death of comrades and by the memory of them, and also in a high degree by the memories of dreams. This is the source of so-called "animism", that is, all those ideas in which the spirits of the dead take the parts of controllers of fortune and bring about either weal or woe in human life. "Fetishism" is a branch of animism, in which the attribute of ability to control fate is carried over to various objects in the environment, such as animals, plants, stones, works of art, especially those that attract the attention on account of their striking character or of some accidental outer circumstance. The phenomena of animism and fetishism are. not only the earliest, but also the most lasting, productions of myth-making apperception. They continue, even after all others are suppressed, in the various forms of superstitions among civilized peoples, such as belief in ghosts, enchantments, charms, etc.

10. After consciousness reaches a more advanced stage personifying apperception begins to deal with the greater natural phenomena which act upon human life both through their changes and through their direct influence such as the clouds, rivers, winds, and greater heavenly bodies. The regularity of certain natural phenomena, such as the alternation of night and day, of winter and summer, the processes in a thunderstorm, etc., gives occasion for the formation of poetical myths, in which a series of interconnected ideas are woven into one united whole. In the way the *nature-myth* arises, which from its very character challenges the poetic power of each individual to

develop it further. It thus becomes gradually a component of popular and then of literary poetry, and undergoes a change in meaning through the fading out of some of the features of the single mythical figures and the appearance of other new features. This change, in turn, makes possible a progressive inner change of the myth, analogous to the change in words, by which it is always accompanied. As the process goes on, single poets and thinkers gain an increasing influence.

In this way, there gradually results a division of the whole content of mythological thought into science (philosophy) and religion, while, at the same time, the nature-gods in religion give place more and more to *ethical* ideas of deity. After this division has taken place, the two departments influence each other mutually in many important ways. Still, these facts must be left to social psychology and the history of civilization, for they must be discussed in the light of special social conditions as well as of general psychological laws.

C. CUSTOMS

11. The development of *customs* is related to that of myths in the same way that outer volitional acts are related to inner motives. Wherever we can trace out the origin of ancient and wide-spread customs with any degree of probability, we find that they are remnants or modifications of certain cult-forms. Thus, the funeral feasts and burial ceremonies of civilized peoples point to a primitive ancestor-worship. Numerous feasts and ceremonies connected with particular days, with the change of the seasons, the tillage of the fields, and the gathering of the harvest, all point back to nature-myths. The custom of greeting, in its various forms betrays its direct derivation from the ceremonies of prayer.

This does not exclude the possibility that other motives, also, especially those of practical utility, have given rise to what were at first individual habits, but gradually. spread throughout a community and thus became laws of custom. The predominant feature of this development, however, is the fact that primitive customs, even when they incidentally serve practical needs, as, for example, the custom of wearing a uniform pattern of clothes, of having meals at a regular time etc., still depend more or less on particular mythological ideas. In fact, it would be hard to think of it as otherwise at a time when consciousness was under the complete control of a myth-making apperception.

12. With customs, as with language, the change in meaning has exercised a modifying influence on their development. As a result of this change, *two* chief kinds of transformation have taken place. In the first, the original mythical motive has been lost and no new one has taken its place. The custom continues as a consequence of associative habit, but loses its imperative character and becomes much weaker in its outward manifestations. In the second class of transformations of a *moral-social*

purpose takes the place of the original mytho-religious motive. The two kinds of change may in any single case be most intimately united; and even when a custom does not serve any particular social end directly, as is the case, for example, with certain rules of deportment, of etiquette, on the manner of dressing, eating, etc., still it may do so indirectly in that the existence of some common rules for the members of a community is favorable to their united life and therefore to their common mental development.

13. The psychological changes in customs as pointed out, constitute the preparation for their differentiation into three spheres, namely those of *custom of law,* and of *morality.* The last two are to be regarded as special forms of custom aiming at moral-social ends. The detailed investigation of the psychological development and differentiation of customs in general is, however, a problem of social psychology, and the discussion of the rise of law and morality belong also to general history and ethics.

14. We have here, in mental communities, and especially in their development of language, myths, and customs, mental interconnections and interactions that differ in essential respects from the interconnection of the psychical compounds in an individual consciousness, but still have just as much reality as the individual consciousness itself. In this sense we may speak of the interconnection of the ideas and feelings of a social community as a *collective consciousness,* and of the common volitional tendencies as a *collective will.* In doing this we are not to forget that these concepts do not mean something that exists apart from the conscious and volitional processes of the individual, any more than the community itself is something besides the union of individuals. Since this union, however, brings forth certain mental products, such as language, myths, and customs, for which only the germs are present in the individual, and since it determines the development of the individual from a very early period, it is just as much an object of psychology as the individual consciousness. For psychology must give an account of the interactions which give rise to the products and attributes of collective consciousness and of the collective will.

14a. The facts arising from the existence of mental communities have only recently come within the pale of psychological investigation. These problems were formerly referred either to the special mental sciences (philology, history, jurisprudence, etc.) or, if of a more general character, to philosophy, that is to metaphysics. If psychology did touch upon them at all, it was dominated, as were the special sciences, history, jurisprudence, etc., by the reflective method of popular psychology, which tends to treat all mental products of communities, to as great an extent as possible, as voluntary inventions aimed from the first at certain utilitarian ends. This view found its chief philosophical expression in the doctrine of a social contract, according to which a mental community is riot something original and natural, but is derived from the voluntary union of a number of individuals. This position is psychologically untenable, and completely helpless in the presence of the problems of social psychology. As one of its after-effects we have even to-day the grossest misunderstandings of the concepts collective consciousness and collective will. Instead of regarding them simply as expressions for the actual agreement and interaction of individuals in a community, some still suspect that there is behind them a mythological being of some kind, or at least a metaphysical substance.

CHAPTER 5

PSYCHICAL CAUSALITY AND ITS LAWS

22. CONCEPT OF MIND

1. Every empirical science has, as its primary and characteristic subject of treatment, certain particular facts of experience whose nature and reciprocal relations it seeks to investigate. In solving these problems it is found to be necessary, if we try not to give up entirely the grouping of the facts under leading heads, to have *general supplementary concepts* that are not contained in experience itself, but are gained by a process of logical treatment of this experience. The most general supplementary concept of this kind that has found its place in all the empirical sciences, is the concept *of causality*. It comes from the necessity of thought. that all our experiences shall be arranged according to reason and consequent, and that we shall remove, by means of second" supplementary concepts and if need be by means of concepts of a hypothetical character, all contradictions that stand in the way of the establishment of a consistent interconnection of this kind. In this sense we may regard all the supplementary concepts that serve for the interpretation of any sphere of experience, as applications of the general principle of causation. They are justified in so far as they are required, or at least rendered probable, by this principle; they are unjustifiable so soon as they prove to be arbitrary fictions resulting from foreign motives, and contributing nothing to the interpretation of experience.

2. In this sense the concept *matter* is a fundamental supplementary concept of natural science. In its most general significance it designates the permanent substratum assumed as existing in universal space, to whose activities we must attribute all natural phenomena. In this most general sense the concept matter is indispensable to every explanation of natural science. The attempt in recent times to raise *energy* to the position of a governing principle, does not succeed in doing away with the concept matter, but merely gives it a different content. This content, however, is given to the concept by means of a second supplementary concept, which relates to the *causal activity* of matter. The concept of matter that has been accepted in natural science up to the present time, is based upon the mechanical physics of Galileo, and uses as its secondary supplementary concept the concept of *force* which is defined as the product of the mass and the momentary acceleration. A physics of energy would have to use everywhere instead of this the concept *energy,* which in the special form of mechanical energy is defined as half the product of the mass multiplied by the square of the velocity. Energy, however, must, just as well as force, have a position in objective space,

184

and under certain particular conditions the points from which energy proceeds may, just as well as the .points from which force proceeds, change their place in space, so that the concept of matter as a substratum contained in space, is retained in both cases. The only difference, and it is indeed an important one, is that when we use the concept force, we presuppose the reducibility of all, natural phenomena to forms of mechanical motion, while when we use the concept of energy, we attribute to matter not only the property of motion without a change in the form of energy, but also the property of the transformability of qualitatively different forms of energy into one another without a change in the quantity of the energy.

3. The concept of *mind* is a supplementary concept of psychology, in the same way that the concept matter is supplementary concept of natural science. It too is indispensable in so far as we need a concept which shall express in a comprehensive way the totality of psychical experiences in an individual consciousness. The particular content of the concept, however, is in this case also entirely dependent on the secondary concepts that give a more detailed definition of psychical causality. In the definition of this content psychology shared at first the fortune of the natural sciences. Both the concept of mind and that of matter arose primarily not so much from the need of explaining experience as from the effort to reach a systematic doctrine of the general interconnection of all things. But while the natural sciences have long since outgrown this mythological stage of speculative definition, and make use of some of the single ideas that originated] at that time, only for the purpose of gaining definite starting-points for a strict methodical definition of their concepts, psychology has continued under the control of the mythological, metaphysical concept of mind down to most modern times, and still remains, in part at least, under its control. This concept is not used as a general supplementary concept that serves primarily to gather together the psychical facts and only secondarily to give a causal interpretation of them but it is employed as a means to satisfy so far as possible the need of a general universal system, including both nature and the individual existence.

4. The *concept of a mind-substance* in its various forms is rooted in this mythological and metaphysical need. In its development there have not been wanting efforts to meet from this position, so far as possible, the demand for a psychological causal explanation, still, such efforts have in all cases been afterthoughts; and it is perfectly obvious that psychological experience alone, independent of all foreign metaphysical motives, would never have led to a concept of mind-substance. This concept has beyond a doubt exercised a harmful influence on the treatment of experience. The view, for example, that all the contents of psychical experience are ideas, and that these ideas are more or less permanent objects, would hardly be comprehensible without such presuppositions. That this concept is really foreign to psychology, is further attested by the close interconnection in which it stands to the concept of material substance. It is regarded either as identical with the latter or else as distinct in nature, but still reducible in its most general formal characteristics to one of the particular forms of the concept matter, namely to the *atom.*

185

5. *Two* forms of the concept mind-substance may be distinguished, corresponding to the two types of metaphysical psychology pointed out above. The one is *materialistic* and regards psychical processes as the activities of matter or of certain material complexes, such as the brain-elements. The other is *spiritualistic* and looks upon these processes as states and changes in an extended and therefore invisible and permanent being of a especially spiritual nature. In this case matter is thought of as made up of similar atoms of a lower order (monistic, or monado-logicial spiritualism), or the mind-atom is regarded as specifically different from matter proper (dualistic spiritualism).

In both its materialistic and spiritualistic forms, the concept mind-substance does nothing for the interpretation of psychological experience. Materialism, does away with psychology entirely and puts in its place an imaginary brain-physiology of the future, or when it tries to give positive theories, falls into doubtful and unreliable hypotheses of cerebral physiology. In thus giving up psychology in any proper sense, this doctrine gives up entirely the attempt to furnish any practical basis for the *mental sciences.* Spiritualism allows psychology is such to continue, but subordinates actual experience to entirely arbitrary metaphysical hypotheses, through which the unprejudiced observation of psychical processes is obstructed. This appears first of all in the incorrect statement of the problem of psychology, with which the metaphysical theories start. They regard inner and outer experience is totally heterogeneous, though in some external way interacting, spheres.

6. It has been shown that the experience dealt with in the natural sciences and in psychology are nothing but components of *one* experience regarded from different points of view: in the natural sciences as an interconnection of objective phenomena and, in consequence of the abstraction from the knowing subject, as *mediate experience;* in psychology as *immediate and underived experience.*

When this relation is once understood, the *concept of a mind-substance* immediately gives place to the *concept of the actuality of mind* as a basis for the comprehension of psychical processes. Since the psychological treatment of experience is supplementary to that of the natural sciences, in that it deals with the immediate reality of experience, it follows naturally that there is no place in psychology for hypothetical supplementary concepts such as are necessary in the natural sciences because of their concept of an object independent of the subject. In this sense, the concept of the actuality of mind does not require any hypothetical determinants to define its particular contents, as the concept of matter does, but quite to the contrary, it excludes such hypothetical elements from the first by defining the nature of mind as the immediate reality of the processes themselves. Still, since one important component of these processes, namely the totality of ideational objects, is at the same time the subject of consideration in the natural sciences, it necessarily follows that substance and actuality are concepts that refer to one and the same general experience with the difference that in each case this experience is looked at from a different point of view. If we abstract from the knowing subject in our treatment of the world of experience, it appears, is a manifold of interacting substances; if, on the contrary, we regard it as the total content of the experience of the subject including the subject itself, it appears as a manifold of interrelated occurrences. In the first case, phenomena are looked upon as *outer phenomena,* in the sense that they would take place just the same, even if the knowing subject were not there at all, so that we may call the form of experience dealt with in the natural sciences outer experience. In the second case, on the contrary, all the contents of experience are regarded as belonging directly to the knowing subject, so that we may call the psychological attitude towards experience that of *inner* experience. In this sense outer and inner experience are identical with mediate and immediate, or with objective and subjective forms of experience. They all serve to designate, not different spheres of experience, but different supplementary points of view in the consideration of an experience which is presented to us as an absolute unity.

7. That the method of treating experience employed in natural science should have reached its maturity before that employed in psychology, is easily comprehensible in view of the practical interest connected with the discovery of regular natural phenomena thought of as independent of the subject; and it was almost unavoidable that this priority of the natural sciences should, for a long time, lead to a confusion of the two points of view. This did really occur as we see by the different psychological substance-concepts. It is for this reason that the reform in the fundamental position of psychology, which looks for the characteristics of this science and for its problems, not in the specifically distinct nature of its sphere, but in its method of considering all the contents presented to us in experience in their immediate reality, unmodified by any hypothetical supplementary concepts - this reform did not originate with psychology itself, but with the *single mental sciences.* The view of mental processes based upon the concept of actuality, was familiar in these sciences long before it was accepted in psychology. This inadmissible difference between the fundamental position of psychology and the mental sciences is what has kept psychology until the present time from fulfilling its mission of serving as a foundation for all the mental sciences.

8. When the concept of actuality is adopted, a question upon which metaphysical systems of psychology have been long divided is immediately disposed of. This is the question of the *relation of body and* mind. So long as body and mind are both regarded as substances, this relation must remain an enigma, however the two concepts of substance may be defined. If they are like substances, then the different contents of experience as dealt with in the natural sciences and in psychology can no longer be understood, and there is no alternative but to deny the independence of one of these forms of knowledge. If they are unlike substances, their connection is a continual miracle. If we start with the theory of the actuality of mind, we recognize the immediate reality of the phenomena in psychological experience. Our physiological concept of the bodily organism, on the other hand, is nothing but a part of this experience, which we gain, just as we do all the other empirical contents of the natural sciences, by assuming the existence of an object independent of the knowing subject. Certain components of mediate experience may correspond to certain components of immediate experience, without its being necessary, for this reason, to reduce the one to the other or to derive one from the other. In fact, such a derivation is absolutely impossible because of the totally different points of view adopted in the two cases. Still, the fact that we have here not different objects of experience, but different points of view in looking at a unitary

experience, renders necessary the existence at every point of relations between the two. At the same time it must be remembered that there is an infinite number of objects that can be approached only immediately, through the method of the natural sciences: here belong all those phenomena that we are not obliged to regard as physiological substrata of psychical processes. On the other hand, there is just as large a number of important facts that are presented only immediately, or in psychological experience: these are all those contents of our subjective consciousness which do not have the character of ideational objects, that is, the character of contents which are directly referred to external objects.

9. As a result of this relation, it follows that there must be a necessary relation between all the facts that belong at the same time to both kinds of experience, to the mediate experience of the natural sciences and to the immediate experience of psychology, for they are nothing but components of a single experience which is merely regarded in the two cases from different points of view. Since these facts belong to both spheres, there must be an elementary process on the physical side, corresponding to every such process on the psychical side. This general principle is known as the principle *of psycho-physical parallelism.* It has an empirico-psychological significance and is thus totally different from certain metaphysical principles that have sometimes been designated by the same name, but in reality have an entirely different meaning. These metaphysical principles are all based on the hypothesis of a psychical substance. They all seek to solve the problem of the interrelation of body and mind, either by assuming *two* real substances with attributes which are different, but parallel in their changes, or by assuming *one* substance with two distinct attributes that correspond in their modifications. In both these cases the metaphysical principle of parallelism is based on the assumption that every physical process has a corresponding psychical process and vice versa; or on the assumption that the mental world is a mirroring of the bodily world, or that the bodily world is an objective realization of the mental. This assumption is, however, entirely indemonstrable and arbitrary, and leads in its psychological application to in intellectualism contradictory to all experience. The psychological principle, on the other hand, as above formulated, starts with the assumption that there is only *one* experience, which, however, as soon as it becomes the subject of scientific analysis, is, in some of its components, open to *two* different kinds of scientific treatment: to a mediate form of treatment, which investigates ideated objects in their objective relations to one another, and to an *immediate* form, which investigates the same objects in their directly known character, and in their relations to all the other contents of the experience of the knowing subject. So far as there are objects to which both these forms of treatment are applicable, the psychological principle of parallelism requires, between the processes on the two sides, a relation at every point. This requirement is justified by the fact that both forms of analysis are in these two cases really analyses of one and the same content of experience, On the other hand, from the very nature of the case, the psychological principle of parallelism can not apply to those contents of experience which are objects of natural-scientific analysis alone, or to those which go to make up the specific character of psychological experience. Among the latter we must reckon the characteristic *combinations* and *relations* of psychical elements and compounds. To be sure, there are combinations of physical processes

running parallel to these, in so far at least as a direct or indirect causal relation must exist between the physical processes whose regular coexistence or succession is indicated by a psychical interconnection, but the characteristic content of the psychical combination can, of course, in no way be a part of the causal relation between the physical processes. Thus, for example, the elements that enter into a spacial or temporal idea, stand in a regular relation of coexistence and succession in their physiological substrata also; or the ideational elements that make up the process of relating or comparing psychical contents, have corresponding combinations of physiological excitation of some kind or other, which are repeated whenever these psychical processes take place. But the physiological processes can not contain anything of that which goes most of all to form the specific nature of spacial [sic] and temporal ideas, or of relating and comparing processes, because natural science purposely abstracts from all that is here concerned. Then, too, there are two concepts that result from the psychical combinations, which, together with their related affective elements, lie entirely outside the sphere of experience to which the principle of parallelism applies. There are the concepts of *value* and *end*. The forms of combination that we see in processes of fusion or in associative and apperceptive processes, as well as the values that they possess is the whole interconnection in of psychical development, can only be understood through *psychological* analysis, in the same way that objective phenomena, such as those of weight, sound, light, heat, etc., or the processes of the nervous system, can be approached only by physical and physiological analysis, that is, analysis that makes use of the supplementary substance-concepts of natural science.

10. Thus, the principle of psycho-physical parallelism in the incontrovertible *empirico-psychological* significance above attributed to it, leads necessarily to the recognition of an *independent psychical causality,* which is related at all points with physical causality and can they come into contradiction with it, but is just as different from it's physical causality as the point of view adopted in psychology, or that of immediate, subjective experience, is different from the point of view taken in the natural sciences, or that of mediate, objective experience due to abstraction. And just as the nature of physical causality can be revealed to us only in the fundamental *laws of nature,* so the only way that we have of accounting for the characteristics of psychical causality is to abstract certain *fundamental laws of psychical phenomena* from the totality of psychical processes. We may distinguish *two* classes of such laws. The laws of one class show themselves primarily in the processes which condition the rise and immediate interaction of the psychical compounds; we call these the *psychological laws of relation.* Those of the second class are derived laws. They consist in the complex effects that are produced by combinations of the laws of relation within more extensive series of psychical facts; these we call the *psychological laws of development.*

23. PSYCHOLOGICAL LAWS OF RELATION

1. There are *three* general psychological laws of relation. We designate them as the laws *of psychical resultants, of psychical relations, and of psychical contrasts.*

2. The *law of psychical resultants* finds its expression in the fact that every psychical compound shows attributes which may indeed be understood from the attributes of its elements after these elements have once been presented, but which are by no means to be looked upon as the mere sum of the attributes of these elements. A compound clang is more in its ideational and affective attributes than merely a sum of single tones. In spacial [sic] and temporal ideas the spacial [sic] and temporal arrangement is conditioned, to be sure, in a perfectly regular way by the cooperation of the elements that make up the idea, but still the arrangement itself can by no means be regarded as a property belonging to the sensational elements themselves. The nativistic theories that assume this implicate themselves in contradictions that cannot be solved; and besides, in so far as they admit subsequent changes in the original space-perceptions and time-perceptions, they are ultimately driven to the assumption of the rise, to some extent at least, of new attributes. Finally, in the apperceptive functions and in the activities of imagination and understanding, this law finds expression in a clearly recognized form. Not only do the elements united by apperceptive synthesis gain, in the aggregate idea that results from their combination, a new significance which they did not have in their isolated state, but what is of still greater importance, the aggregate idea itself is a new psychical content that was made possible, to be sure, by these elements, but was by no means contained in them. This appears most strikingly in the more complex productions of apperceptive synthesis, as, for example, in a work of art or a train of logical thought.

3. The law of psychical resultants which expresses a principle which we may designate, in view of its results, as a *principle of creative synthesis.* This has long been recognized in the case of higher mental creations, but generally not applied to the other psychical processes. In fact, through an unjustifiable confusion with the laws of physical causality, it has even been completely reversed. A similar confusion is responsible for the notion that there is a contradiction between the principle of creative synthesis in the mental world and the general laws of the natural world, especially that of the conservation of energy. Such a contradiction is impossible from the outset because the points of view for judgment, and therefore for measurements wherever such are made, are different in the two cases, and must be different, since natural science and psychology deal, not with different contents of experience, but with one and the same content viewed from different sides. Physical measurements have to do with *objective masses, forces, and energies.* These are supplementary concepts which we are obliged to use in judging objective experience; and their general laws, derived as they are from experience, must not be contradicted by any single case of experience. Psychical measurements, which are concerned with the comparison of psychical components and their resultants, have to do with *subjective values and ends.* The subjective value of the

whole may increase in comparison with that of its components; its purpose may be different and higher than theirs without any change in the masses, forces, and energies concerned. The muscular movements of an external volitional act, the physical processes that accompany sense-perception, association, and apperception, will follow invariably the principle of the conservation of energy. But the mental values and ends that these energies represent may be very different in quantity even while the quantity of these energies remains the same.

4. The differences pointed out show that *physical* measurement deals with *quantitative values,* that is, with quantities that admit of a variation in value only in the one relation of the quantity of the phenomena measured. *Psychical* measurement on the other hand, deals in the last instance in every case with *qualitative values,* that is, values that vary in degree only in respect to their qualitative character. The ability to produce purely *quantitative* effects, which we designate as *physical energy* is, accordingly, to be clearly distinguished from the ability to produce *qualitative* effects, or the ability to produce values, which we designate as *psychical energy.*

On this basis we can not only reconcile the *increase of psychical energy* with the *constancy of physical energy* as accepted in the natural sciences, but we find in the two reciprocally supplementary standards for the judgment of our total experience.. The increase of psychical energy is not seen in its right light until it is recognized as the reverse, subjective side of physical constancy. The former, being as it is indefinite, since the measure may be very different under different conditions, holds only *under the condition that the psychical processes are continuous.* As the psychological correlate of this increase we have the fact which forces itself upon us in experience, that *psychical values disappear.*

5. The *laws, of psychical relations* supplements that of resultants; it refers not to the relation of the components of a psychical interconnection to the value of the whole, but rather to their reciprocal relation. The law of resultants thus holds for the synthetic processes of consciousness, the law of relations for the analytic. Every resolution of a conscious content into its single members is an act of relating analysis.

Such a resolution takes place in the successive apperception of the parts of a whole which is ideated at first only in a general way, a process which is to be seen in sense-perceptions and associations, and then in clearly recognized form in the division of aggregate ideas. In the same way, every apperception is an analytic process whose two factors are the emphasizing of one single content and the marking off of this one content from all others. The first of these two partial processes is what produces *clearness,* the second is what produces *distinctness* of apperception. The most complete expression of this law is to be found in the processes *of apperceptive analysis* and the simple *relating* and *comparing* functions upon which it is based. In the latter more especially, we see that the essential content of the law of relations is the principle that every single psychical content receives its significance from the relations in which it stands to other psychical contents. When these relations are *quantitative,* this principle takes the form of a principle of *relative quantitative comparison* such as is expressed in *Weber's law.*

6. The *law of psychical contrasts* is, in turn, supplementary to the law of relations. It refers, like the latter, to the relations of psychical contents to one another., It is itself based on the fundamental division of the immediate contents of experience into objective and subjective components, a division which is due to the very conditions of psychical development. Under subjective components are included all the elements and combinations of elements which, like the feelings and emotions are essential constituents of *volitional processes.* These are all arranged in *groups* made up of *opposite qualities* corresponding to the chief affective directions of pleasurable and unpleasurable, exciting and depressing, straining and relaxing feelings. These *opposites* obey in their succession the general *law of intensification through contrast* In its concrete application, this law is always determined in part by special temporal conditions, for every subjective state requires a certain period for its development; and if, when it has once reached its maximum, it continues for a long time, it loses its ability to arouse the contrast-effect. This fact is connected with the other, that there is a certain medium, though greatly varying, rate of psychical processes most favorable for the intensity of all feelings and emotions.

This law of contrast has its origin in the attributes of the subjective contents of experience, but is secondarily applied to the ideas and their elements also, for these ideas are always accompanied by more or less emphatic feelings due either to their own content onto the character of their spacial [sic] and temporal combination. Thus the principle of intensification through contrast finds its broader application especially in the case of certain sensations, such as those of sight, and in the case of spacial [sic] and temporal ideas.

7. The law of contrast stands in close relation to the two preceding laws. On the one hand, it may be regarded as the application of the general law of relations to the special case where the related psychical contents range between opposites. On the other hand, the fact that under suitable circumstances antithetical psychical processes may intensify each other, while falling under the law- of contrast, is at the same time a special application of the principle of creative synthesis.

24. PSYCHOLOGICAL LAWS OF DEVELOPMENT

1. We have as many psychological laws of development as we had laws of relation, and the former may be regarded as the application of the latter to more comprehensive psychical interconnections. We designate the laws in question as those of mental *growth of heterogony of ends,* and of *development, towards opposites.*

2. The *law of mental growth* is as little applicable to all contents of psychical experience as any other psychological law of development. It holds only under the limiting condition under which the law of resultants, whose application it -is, holds, namely under the condition of the *continuity of the processes*. But since the circumstances that tend to prevent the realization of this condition, are, of course, much more frequent when the mental developments concerned include a greater number of psychical syntheses, than they axe in the single syntheses themselves, it follows that the law of mental growth can be demonstrated only for certain developments taking place under normal conditions, and even here only within certain limits. Within these limits, however, the more comprehensive developments, as, for example, the mental development of the normal individual and the development of mental communities, are obviously the best exemplifications of the fundamental law of resultants which lies at the basis of this development.

3. The *law of heterogony of ends* is most closely connected with the law of relations, but it is also based on the law of resultants, which is always to be taken into consideration when dealing with the larger interconnections of psychical development. In fact, we may regard this law as a principle of development which controls the changes arising, as results of successive creative syntheses, in the relations between the single partial contents of psychical compounds. The resultants arising from united psychical processes include contents that were not present in the components, and these new contents may in turn enter into relation with the old components thus changing again the relations between these old components and consequently the new resultants that arise from them. This principle of continually changing relations is most striking when an *idea of ends* is formed on the basis of the given relations. Here the relation of the single factors to one another is regarded as an interconnection Of means which has for the end aimed at, the product arising from the interconnection. The relation between the actual *effects* in such a case and the ideated ends is such that secondary effects always arise that were not thought of in the first ideas of end. These new effects enter into new series of motives, and thus modify the old ends or add new ones to them.

The principle of heterogony of ends in its broadest sense dominates all psychical processes. In the special teleological coloring which has given it its name, however, it is to be found primarily in the sphere of *volitional processes,* for here the ideas of end attended by their affective, motives are of the chief importance. In the various spheres of applied psychology it is therefore especially *ethics* for which this law is of great importance.

4. The *law of development towards opposites* is an application of the law of intensification through contrast, to more comprehensive interconnections which form in themselves series of developments. These series, in accordance with the fundamental law of contrasts, are of such a character that feelings and impulses which were of small intensity at first, increase gradually in intensity through contrast with feelings of opposite quality that were for a time predominant, until, finally, they gain the ascendancy over the formerly predominant feelings and are themselves for a longer or shorter time in control. From this point the same alternation may be once or even

several times repeated. But generally the principles of mental growth and heterogony of ends operate in the case of such an oscillation, so that succeeding phases are like corresponding antecedent phases in their general affective direction, but still essentially different in their special components.

The law of development towards opposites shows itself in the mental development of the individual, partly in a purely individual way within shorter periods of time, and partly in certain universal regularities in the relation of various periods of life. It has long been recognized that the predominating temperaments of different periods of life present certain contrasts. Thus, the light, sanguine excitability of childhood, which is seldom more than superficial, is followed by the slower but more retentive temperament of youth with its frequent touch of melancholy. Then comes manhood with its mature character, generally quick and active in decision and execution, and last of all, old age with its leaning toward contemplative quiet. Even more than in the individual does this principle of antithesis find expression in the alternation of mental tendencies that appear in the social and historical life of communities, and in the reactions of these tendencies on civilization and customs and on social and political development. In the same way that the principle of heterogony of ends applied chiefly to the domain of moral life, this principle of development towards opposites finds its chief significance in the more general sphere of *historical* life.